The 13th Element

ment

The Sordid Tale of Murder, Fire, and Phosphorus

John Emsley

John Wiley & Sons, Inc.

Copyright © 2000 by John Emsley. All rights reserved.

Published by John Wiley & Sons, Inc., New York
Published simultaneously in Canada

First published in Great Britain with the title
The Shocking History of Phosphorus
by Macmillan, an imprint of Macmillan Publishers Ltd., in 2000.

'Pack Up Your Troubles (In Your Old Kit Bag)'
Words by George Asaf and Music by Felix Powell © 1915,
reproduced by permission of Francis Day & Hunter Ltd, WC2H 0EA, England.

This publication is designed to provide accurate and authoritative information in
regard to the subject matter covered. It is sold with the understanding that the
publisher is not engaged in rendering professional services. If professional advice or
other expert assistance is required, the services of a competent professional person
should be sought.

ISBN 0-471-44149-X

Printed in the United States of America

10 9 8 7

Contents

Acknowledgments

My interest in the history of phosphorus began many years ago when I collaborated on writing *The Chemistry of Phosphorus* with my colleague Dennis Hall. While collecting data for that book I came across all kinds of human-interest stories which were not suitable for a textbook of chemistry, but were ideal for a popular science article, and this appeared in the magazine *New Scientist*. An unexpected response to this was an invitation to give a talk – with demonstrations – for the Molecule Theatre, as part of a series of public lectures organized by Sir Bernard and Lady Miles. Following that came further invitations to give the talk and at many of these I would be approached by a member of the audience with yet another story to add to my collection of phosphorus anecdotes.

In 1992, I was invited to turn the talk into a script for a radio programme, *The Shocking History of Phosphorus*, which was produced by Louise Dalziel of BBC Radio Scotland and broadcast on BBC Radio 4 in September of that year. The programme won a Glaxo Award for popular science broadcasting, and a Sony Award for memorable radio programmes. As a result I accumulated yet more material from listeners.

It would be impossible now to name all the people who have contributed to the story of phosphorus over the years, but some deserve special mention, especially those who have checked chapters of *The Shocking History of Phosphorus* dealing with areas in which they are acknowledged specialists, and suggested ways to improve them. They were: Tim Jones and Steve Marsden (Chapter 1), Fred Holmes (Chapter 2), Peter Fell (Chapter 3), Nik Wachsmann

(Chapter 7), Guy Selby-Lowndes (Chapter 8), Bill Duley (Chapters 10, 11 and 13) and Ed Rider (Chapter 13). Special appreciation goes to Julian Emsley, Mari Evans and Patrick Walsh for their encouragement and support.

I should also like to thank the late Bill Owen MBE, for providing the lyrics of his musical *The Matchgirls* (Chapter 6), Michael Frayn for permission to quote from his play *Copenhagen* and Len Deighton for permission to quote from his book *Bomber* (both in Chapter 7).

The 13th Element

Introduction

The Shocking History of Phosphorus is the first biography of a chemical element, told through the stories of a rich tableau of characters who were involved with it during its 300-year history of curious, bizarre and horrific events. Phosphorus was discovered by the alchemists, researched by the early chemists, exploited by the industrialists of the nineteenth century and abused by the combatants of the twentieth. Its capacity for evil cursed all who tried to exploit it, from the would-be murderer to the worldwide manufacturer. But set against this tale of woe are a few remarkable benefits that phosphorus brought and it is an important ingredient in many of the things we use in our everyday lives. Still it continues to surprise, as it did in the 1990s when it was shown to be the likely cause of mysterious lights, graveyard ghosts and possibly even of spontaneous human combustion.

Before you begin its story, you need to know a little about its name. The word **phosphorus** is derived from the ancient Greek words *phos*, meaning 'light', and *phorus*, meaning 'bringing'. Phosphorus is the name of the chemical element, and its chemical symbol is P. It does not occur naturally, only in the oxidized form of **phosphates**, which are minerals in which each phosphorus atom is surrounded by four oxygens. **Organophosphate** is the name given to a phosphate molecule that has organic groups attached, by which is meant derivatives of carbon. DNA is an organophosphate, and so are nerve gases and insecticides.

In everyday speech the words phosphorus and phosphate are

often used interchangeably, so that we talk of phosphorus fertilizers and phosphorus pollution, when we really mean phosphate
fertilizers and phosphate pollution. The phosphorus cycle in
Nature governs all life on Earth and this too refers to phosphate
or organophosphate. Whenever possible I shall use the correct
term in *The Shocking History of Phosphorus*, but sometimes I
will follow convention and talk of phosphorus when this is the
term in general use. If there is any ambiguity I will use the term
elemental phosphorus to make it clear that what is being referred
to is phosphorus itself and not phosphate.

1. Out of alchemy

While thus employed, Gerard was busy about the seated corpse and to his amazement, Denys saw a luminous glow spreading rapidly over the white face. Gerard blew out the candle. And on this the corpse's face shone still more like a glowworm's head. Denys shook in his shoes and his teeth chattered.

'What in Heaven's name is this?' he whispered.

'Hush! 'tis but phosphorus. But 'twill serve.'

In half a minute Gerard's brush made the dead head a sight to strike any man with dismay. He put his art to a strange use and one unparalleled perhaps in the history of mankind. He illuminated his dead enemy's face to frighten his living foe; the staring eyes he made globes of fire; the teeth he left white ... but the palate and tongue he tipped with fire ... and on the brow he wrote in burning letters LA MORT.

Charles Reade's best-selling novel, *The Cloister and the Hearth*, from which this extract is taken, was published in 1861. The story recounts the adventures of two friends, Gerard and Denys, as they romp their way across Europe in the middle years of the fifteenth century. The book was much admired, not only for being a good read, but for its attention to historical detail, although the author got it very wrong when it came to phosphorus. This element was unknown in the Middle Ages, even if it was an everyday item when Reade wrote his book.

The time-slip is partly excusable because the origin of phosphorus was shrouded in mystery. It had even been suggested that it was known in the days of ancient Rome, and it may well be that the secret of its manufacture was discovered – and lost – more than once down the centuries. This we can attribute to the paranoid secrecy of the alchemists as they searched endlessly for the philosopher's stone, the mythical compound that would turn base metals like lead into gold. It was one of their number who finally revealed phosphorus to the world, but that was 200 years after the time of Gerard and Denys.

Uncertainty still surrounds the date on which phosphorus was first made. We can be fairly sure the place was Hamburg in Germany, and that the year was probably 1669, but the month and day are not recorded, though it must have been night-time. The alchemist who made the discovery stumbled upon a material the like of which had never been seen. Unwittingly he unleashed upon an unsuspecting world one of the most dangerous materials ever to have been made.

On that dark night our lone alchemist was having no luck with his latest experiments to find the philosopher's stone. Like many before him he had been investigating the golden stream, urine, and he was heating the residues from this which he had boiled down to a dry solid. He stoked his small furnace with more charcoal and pumped the bellows until his retort glowed red hot. Suddenly something strange began to happen. Glowing fumes filled the vessel and from the end of the retort dripped a shining liquid that burst into flames. Its pungent, garlic-like smell filled his chamber. When he caught the liquid in a glass vessel and stoppered it he saw that it solidified but continued to gleam with an eerie pale-green light and waves of flame seemed to lick its surface. Fascinated, he watched it more closely, expecting this curious cold fire to go out, but it continued to shine undiminished hour after hour. Here was magic indeed. Here was phosphorus.

That Hamburg alchemist was Hennig Brandt, a rather pompous man who insisted on being called *Herr Doktor* Brandt.

His first thought was that he had at last found what he had been searching for. Surely this wondrous new material was the philosopher's stone? If so, he had better keep the secret to himself while he made his fortune. And so for six long years he hid his discovery from the world, until his wealth was all but spent. Magical though it was, his phosphorus stubbornly refused to bring the riches of which he dreamed. Today we know his was a vain hope, but the luminescent material he had made was to create great wealth for a few in the centuries ahead. It was to create untold misery for many more.

Frustrated by his inability to make gold with phosphorus, Brandt finally revealed its existence to friends and neighbours and soon it was the talk of Hamburg. While Brandt enjoyed the fame it brought him he still kept secret the method of its making, although he let it be known that it was of human origin. A few of those he showed it to became fascinated by it and, guessing that it must derive from urine, tried to make it for themselves. They sometimes succeeded and when they wrote about it they were less coy about how it was made, which is why its discovery came to be attributed to others. Brandt never published an account of his discovery and he soon faded into obscurity. None of his papers has survived, but, even if they had, it is most unlikely that we would be able to interpret their alchemical symbols and arcane language. Those who dabbled in that ancient lore guarded their secrets well, fearful lest their work – or their chicanery – should be betrayed to the princes, kings and emperors who funded them. Brandt, who was arguably the most successful of them all, was funded by his first and second wives.

There is no recorded date of Brandt's birth, although it was probably in the 1620s, nor are there records of the year he married his first wife, or of when she died. We know they had at least two children. Then he married Margaretha, a rich widow who already had a son by her first marriage, and this stepson helped Brandt in his laboratory. It was her wealth that enabled the alchemist to continue his search for the philosopher's stone and thereby discover phosphorus.

The little we do know about Brandt comes from letters written by Margaretha and from comments of those he had dealings with. We know he lived in the Michaelisplatz part of Hamburg at the time he made his discovery and was probably a native of that city. He was said by some to be a man of humble birth who in his youth had been apprenticed to a glass-maker, a trade which would stand him in good stead when he turned his attention to alchemy because it enabled him to produce the high temperatures necessary to generate phosphorus. However, in one letter written by Margaretha in November 1678 she said her husband was of high social standing, which might well have been true.

In his early days Brandt had been a soldier when such men were in great demand* and even held the rank of a junior officer, which perhaps indicated that he was from a respectable family. His first wife came with a substantial dowry and it was her money that allowed him to take up alchemy when he left the army. By the time she died he had spent most of it on his fruitless search, but his depleted coffers were soon replenished by his second marriage.

Those who knew Brandt have also left a few clues to his character. He was said to be of whimsical temperament and secretive, but Ambrose Godfrey, who sought him out around 1680, hoping to discover how phosphorus was made, referred to him as 'old honest Brandt of Hamburg', although others doubted

---

* This was the time of the so-called Thirty Years' War (1618–48), which raged across Europe and pitted Catholics against Protestants, Austria against the German states, the German states against one another, and France against Spain. Meanwhile Denmark and Sweden fought for control of the countries surrounding the Baltic, and in Britain there was a civil war between royalists and parliamentarians.

his honesty and Johann Kunckel, who came earlier on the same mission, often mocked him, calling into question his right to call himself *Herr Doktor* Brandt. Kunckel was particularly bitter because of the way that Brandt had tricked him, yet he was to have the satisfaction of independently discovering how to make phosphorus and it was Kunckel who was often cited as *the* discoverer in the next century. Brandt revealed the existence of phosphorus in 1675, but in the years to follow it was other alchemists who basked in the fame of its awesome glow.

1675

Kunckel was born in 1630, the son of the court alchemist of Duke Frederick of Holstein. Like his father, he was attracted to alchemy and as a young man he obtained a position as alchemist to Duke Franz Carl of Sachsen-Lauenburg. In 1667, he moved to the Court of John George II, Elector of Saxony, but his post there was terminated in 1675 when it became clear that he was unable to turn lead into gold. Such was the esteem in which he was held, however, that he was offered a post teaching 'chemistry'* at the University of Wittenberg and it was there, in his laboratory, that Kunckel started investigating materials that shone in the dark.

At the time, a few materials were known that were luminescent, such as rotting wood and decaying fish, crystals of fluorspar (calcium fluoride) and Bologna stone (made from naturally occurring barium sulphate). This last mineral was first reported around 1640 by Vincenzio Casciorolo, an alchemist of Bologna. If crystals of Bologna stone were left in the sun during the day, they would glow for several hours during the night. The English

---

* Chemistry then was more akin to alchemy, and is not recognizable as the science that it was to become in the following century.

diarist John Evelyn saw some on his travels in Italy in May 1645 and marvelled at their blue radiance.

Kunckel's interest in luminescence was heightened when he learned from a visitor that Hamburg was abuzz with talk of a new material that far outshone any other, including Bologna stone. Not only did this material glow in the dark for hour after hour, day after day, it did not need to be exposed to the sun in order to do so. Kunckel decided to seek it out, and he had not been long in that city before he learned of the man who possessed it: *Herr Doktor* Brandt. He called on Brandt, who was clearly flattered by his academic visitor and agreed to show him his phosphorus, although, as always, he was reticent about how he had made it. During the visit one of Brandt's children fell and grazed his face and, suspecting Brandt's title of *Doktor* was not genuine, Kunckel advised him to apply *oleum cerae* (oil of wax) to the child's cuts, whereupon Brandt revealed his ignorance by not recognizing this common medicament by its Latin name.

Even if he was not a medical man, Brandt was certainly an accomplished alchemist and there was no doubting that he had discovered a phosphorus to outshine all others. Kunckel was fascinated by what he saw and may even have offered to buy some of it to take back to Wittenberg to study. Brandt could well have entertained this idea because we know that he was again short of money, having spent most of his second wife's assets on his alchemical experiments. However, he was not prepared to sell the recipe for making it, which is what Kunckel really wanted.

Excited by what he had seen, Kunckel immediately wrote to a fellow alchemist, fifty-year-old Daniel Kraft, who lived in Dresden, describing the remarkable new phosphorus. Kraft immediately, and without telling Kunckel, rushed to Hamburg to see it for himself. He realized that this was a wonder indeed and, though he was sceptical about it being the philosopher's stone, he saw that it might well make gold in the right hands – his. He asked Brandt if he was willing to sell him some of his phosphorus. Indeed he was. In fact he would sell Kraft all that he had – and all that he could make in the future. Brandt could hardly believe his good fortune.

As Brandt and Kraft were talking there came a knock at the door, and when Brandt answered it he found Kunckel standing there. Quickly he stepped outside and closed the door behind him, saying that his wife was sick and someone was with her. Kunckel expressed a desire to know more about phosphorus and wanted to purchase some, but Brandt was suddenly reluctant to sell him any, saying he had none to spare and claiming that when he had recently tried to make more phosphorus he had had no success. When Kunckel again pressed him to say how it could be produced, Brandt, desperate to get rid of his unwanted visitor, admitted that it came from urine, but that was all he would say. He bade Kunckel farewell and went back indoors.

There Brandt continued his negotiations with Kraft and signed a bond for a fee of 200 thalers (about £4,000 or $6,000 in today's money), which gives us some indication of the dire financial straits in which Brandt now found himself. A few years earlier he and his wife had enjoyed an income of 1,000 thalers per year. Now he gave Kraft all the phosphorus he had and said he would make him more as required. Part of the deal was that Brandt should not talk about phosphorus to anyone and Kraft stressed that in particular he must never tell Kunckel how it was made. Brandt agreed.

The following day, Kunckel was surprised to meet Kraft in the street and asked why he had come to Hamburg. Kraft admitted that he had been to see the new phosphorus but said that he too had found Brandt difficult to deal with and that he could get nothing out of him about how it was made. On that note they parted and Kunckel returned to Wittenberg, where he began to experiment with urine. After several unsuccessful attempts to convert it to phosphorus he wrote to Brandt asking again how to make it, but Brandt's letter in reply said nothing, although he did reveal that he was now under an obligation to Kraft not to reveal the secret to anyone. Kunckel realized he had been tricked. He became all the more determined to discover how to make it himself and he soon succeeded.

1676

In the spring of 1676 Kraft put into operation his plan to make himself rich from his exclusive possession of the new phosphorus. He devised a series of dramatic demonstrations that he could perform with it and set out to travel the courts of Europe, where, for a fee, he would entertain royalty and courtiers with its remarkable properties. Like Brandt and Kunckel, Kraft was also an alchemist, although as a student he had graduated as an MD and, for a time, was employed as a physician by a mining company.

For the next two years Kraft displayed his phosphorus to most of the crowned heads of Europe. He even boasted to those who attended his demonstrations that he was the discoverer, and we know of this from the writings of one, Johann Elsholtz, who was invited to the display put on for the Elector of Brandenburg on the evening of 24 April 1676. At nine o'clock all the candles were extinguished and Kraft gave a demonstration of 'perpetual fire' for the benefit of the Court. Elsholtz came away from the event believing that Kraft was the discoverer of phosphorus and wrote as much when he published an account of what he had witnessed in a philosophical journal.

The French chemist and apothecary Nicholas Leméry also saw a demonstration by Kraft, probably at Paris, and he too said in his textbook of 1683, *Cours de Chymie*, that Kraft was the discoverer. Leméry had been particularly flattered when Kraft gave him a little of the new material and he even managed to cause an accident with it:

After some experiments made one day at my house upon the phosphorus, a little piece of it being left negligently upon the table in my chamber, the maid making the bed took it up in the bedclothes she had put upon the table, not seeing the little piece. The person who lay afterwards in the bed, waking at night and feeling more than ordinary heat, perceived that the coverlet was on fire.

Despite many failures, Kunckel had persisted with his experiments with urine and, in April 1676, he finally succeeded in producing a sample, a yellow blob in a sticky black mass. It hardly compared with Brandt's phosphorus, but Kunckel now felt confident enough to write again to Brandt suggesting they compare notes about their respective methods. Suspecting that this was merely a ruse to get him to reveal his recipe, Brandt replied that he was reluctant to send details via the post in case others should discover the secret. Kunckel suggested that he write in alchemical code, but no reply was forthcoming from Hamburg before 25 July, when Kunckel made a much better specimen of phosphorus. From that point on he realized he had no further need of Brandt. Brandt finally wrote, offering to sell the secret of his method, but Kunckel simply spurned the offer. Kunckel also made gold from his phosphorus by selling it as raw material for medicines, as we shall see in Chapter 3, and he too was invited to the courts of Saxony and Brandenburg to display his discovery.

Eventually Kunckel published a paper on the properties of phosphorus – although even he did not reveal how it could be made. By now he had become fascinated by its source, reasoning that what came out of the human body must also go into it. He investigated various foods and was amazed to discover that many things of plant or animal origin, when strongly heated in a furnace, were capable of producing phosphorus. He boasted that he could obtain phosphorus from all that God had created:

mammals, fish, birds, plants and trees. It was a remarkable observation, but he was right.

Kunckel's interest in phosphorus waned as his career developed. He was even quoted as saying: 'I am not making it [phosphorus] any more, for much harm can come of it.' In 1679 he became Director of the Alchemical Laboratory of Frederick William, Duke of Brandenburg, in Berlin and ten years later had risen to become Minister of Mines for all of Sweden, under Charles XI, where he was eventually honoured with the title Baron von Löwenstern. Kunckel died in 1703 but his fame as a chemist and his links with phosphorus were to endure. More than a century later, in his *Dictionary of Practical and Theoretic Chemistry*, published in 1808, William Nicholson referred to the element as Kunckel's phosphorus, calling it, so he thought, after its discoverer. By then Brandt's eclipse was total. What kept his name alive, and eventually regained him the title of the first person to make phosphorus, was the interest of Gottfried Wilhelm Leibnitz.*

1677

Kraft continued making his progress among the royal palaces of Europe, enjoying the fame and glory that his phosphorus brought. He never revealed the source of the new material and never mentioned Brandt. And so things might have continued had Kraft not gone to Hanover in 1677, there to perform before Duke Johann Frederick of Saxony. In his audience was thirty-

---

* Leibnitz is today remembered as the inventor of infinitesimal calculus, which transformed mathematics, but he was conversant with all branches of contemporary science including optics, mechanics, statistics, logic and probability theory. When he died at Hanover in 1716 most of his work was still unpublished.

one-year-old Leibnitz, who was employed as historian and librarian to the Duke. He was intrigued by the samples of phosphorus he saw and wanted to know more about it. How was it made? Kraft, of course, would not say.

A few months later, Leibnitz made a trip to Hamburg to buy books and there he learned, quite by chance, that a local man, Brandt, was also able to make the new phosphorus. Leibnitz searched him out and confirmed that this was so. Brandt, who by now was short of money again, was more than willing to explain that he, not Kraft, was the discoverer and that Kraft was merely exploiting him. In fact Brandt was by now refusing to supply Kraft with any more phosphorus, saying that all his attempts to produce it were proving unsuccessful. Leibnitz asked Brandt how he made phosphorus and Brandt not only freely admitted he got it from urine, he even offered to show his visitor how to make it – for a fee, of course. Leibnitz replied that he would enquire of Duke Frederick whether Brandt could come to Hanover to do this, and on that hopeful note they parted.

Brandt really was in need of money. Although he found reasons not to supply Kraft with any more phosphorus, it hardly mattered to Kraft who now knew that Kunckel had found a way to make it. In the spring of 1677, Kunckel had written to him offering to supply him with it but Kraft replied that he preferred Brandt's phosphorus as it was of better quality. Not that Kraft needed any more phosphorus, having almost exhausted the list of clients willing to view it. However, there remained one royal court Kraft had not yet visited where he knew he would be made most welcome and that was the Court of St James's in London. What transpired on that memorable visit we shall see in the next chapter.

After his visit to Hamburg, Leibnitz had returned to Hanover, where he suggested to the Duke that Brandt might be employed by the Court as the resident alchemist. The Duke agreed, saying he could be employed – at 10 thalers per month – and he accepted Leibnitz's suggestion that Brandt should receive six months' pay in advance if he revealed how phosphorus was made. On 18 July 1677, a contract was drawn up and sent to Brandt, which presented him with a rather awkward problem. Rebuffed by Kraft, Brandt let it be known that he was available for employment as an alchemist and one who already knew how to make the phosphorus. He had even entered into negotiations with a view to becoming resident alchemist for the Duke of Mecklenburg-Güstrow. He wrote to Leibnitz telling him of this, hinting that he was about to accept the post although nothing definite had been agreed.

Leibnitz persuaded Brandt to visit Hanover before deciding which offer to accept. On his return to Hamburg, Brandt made a final plea to Kraft, reminding him of their agreement and asking for more money, but Kraft was no longer interested and advised him to take the post at Hanover. Brandt's wife Margaretha wrote to Leibnitz on her husband's behalf asking for a higher salary, but to no effect. In the end, Brandt accepted the offer from Hanover, telling Leibnitz that he would need access to a small furnace and copious quantities of urine.

1678

Eventually, in the summer of 1678, Brandt journeyed to Hanover, and there outside the city he set up his apparatus, and with Leibnitz's help he made some phosphorus. Leibnitz was duly impressed with its quality and properties which were exactly those he had witnessed when Kraft had put on his demonstrations for the Duke. When he asked Brandt how he had discovered phosphorus, Brandt told him that he had first made it in 1669, and said that he got the idea from a book by F. T. Kessler of Strasbourg called *400 Auserlesene Chemische Process* ('400 Selected Chemical Processes'), published in 1630, in which there was a recipe for turning base metals into silver using concentrated urine, alum and saltpetre.*

Brandt returned to Hamburg, promising to make more phosphorus there and send it to Leibnitz, but nothing was forthcoming. Leibnitz wrote to Brandt a few times, querying his silence, until on 11 December 1678 he finally received a reply from the alchemist who claimed that he had been unable to work because he and two of his children had been ill and that his eldest daughter had died. Brandt asked Leibnitz for more money, and after a further exchange of letters and an increased allowance of 10 thalers per week he agreed to return to Hanover and make more phosphorus. Again the experiment was a success, so much so that Leibnitz now published an account of how to make phosphorus from urine, although he omitted to mention the debt he owed to Brandt. It was not that Leibnitz wanted to deny the tiresome alchemist his rightful claim to be the discoverer: whenever he was asked about it he would admit that Brandt had shown him how to make it.

Perhaps not surprisingly, since Brandt's name was missing from all the early accounts of phosphorus, his contribution was overlooked by succeeding generations of scientists, and he was soon forgotten. About what happened to Brandt after his second trip to Hanover we know virtually nothing. He lived to be at least sixty because he was still alive in 1692, according to Leibnitz, and even in 1710 Leibnitz said that he had not heard of his death, which could mean that he lived to be eighty or more. If so, Brandt must have been aware that by then his discovery had fostered a new industry in London, and a vogue for taking the element as a medicine. What rescued Brandt from oblivion was Leibnitz's papers, among which were the letters from Margaretha.

---

* Alum is potassium aluminium sulphate, and saltpetre is potassium nitrate. Needless to say the recipe rarely, if ever, succeeded.

Phosphorus – the unanswerable questions

Whatever Brandt's fate, his place in the story of chemistry was assured by his fame, or notoriety, in being the first person to have discovered phosphorus. The shocking history of phosphorus had begun. From our vantage point of the twenty-first century we can answer the questions these early investigators must have asked themselves: What is phosphorus? Why does it glow in the dark? How can it be produced from urine?

The first is easy to answer: phosphorus is a chemical element. Its properties are now well known, its forms varied and its chemistry well understood – see Appendix below, p. 303.

The second question proved unanswerable for more than 300 years. The famous painting *The Alchymist* by Joseph Wright of Derby (1734–97) – Plate 1 – captures the wonder of the discovery but exaggerates the intensity of the light that it emits. Even so, the glow from phosphorus is impressive and, although other ways of generating light by chemical reactions are now known (see box), the length of time that phosphorus will glow is still remarkable.

It was not until 1974 that the correct explanation for phosphorus's luminescence was forthcoming. It was revealed by R. J. van Zee and A. U. Khan in the *Journal of the American Chemical Society* (p. 6805). The explanation lies in the slow chemical reaction between phosphorus and the oxygen of the air which takes place on the surface of phosphorus, forming two species which have only a fleeting existence: a molecule of formula HPO and an oxide of formula P_2O_2. Both emit visible light. Very little of these unstable species need be formed to produce the luminescence, which is why a piece of phosphorus in a closed vessel continues to glow for hours and days, until the last trace of oxygen has been used up.

How can phosphorus be produced from urine? This waste

Living things that give off light

Although the words phosphorus and phosphorescence derive from the same Greek word, phosphorus's glow is not due to phosphorescence. Phosphorescence is the process whereby light is first absorbed by a body and then re-emitted from it some time later; Bologna stone is phosphorescent.

Phosphorus itself is *luminescent*. Luminescence describes the process in which light is emitted as a result of an energy change within a substance, and there are many living things that can luminesce, such as flies, fungi, fish and jellyfish. (The technical term for this is *bio*luminescence.) Even some bacteria are able to emit light, which accounts for the luminescence that occurs in the wake of a boat at sea, or the glow of decaying wood and fish. All produce a chemical, luciferin, which is a sulphur–nitrogen compound, and an enzyme, luciferase, which enables the luciferin to react with oxygen and as it does so it releases some of its energy in the form of visible light. The genes responsible for bioluminescence have been identified in the light-emitting bacterium *Photorhabdus luminescens*, and research is now going on into transferring these into other organisms to act as biological sensors.

material remained the only source of the element for nigh on a century, and it is easy to understand why: phosphate is one of the most abundant components of urine because human beings take in much more phosphate from their diet than they actually need. Most of the excess phosphorus is passed in the urine as phosphate, at the rate of around 1.4 g phosphorus per day.* The main chemicals in urine are given in Table 1.1.

---

* The body also loses phosphorus through the faeces, but less is excreted

Table 1.1: The chief components in urine
(grams per litre)

Constituent	Adult male	Adult female
Creatine	52	92
Urea	21	21
Chloride	6.5	4.7
Sodium	4.0	2.9
Potassium	2.2	1.8
Amino acids*	ca 2.3	ca 2.3
Phosphorus	1.4	1.4
Ammonia	0.68	0.51
Magnesium	0.26	0.21

* Chiefly glycine, histidine and taurine.

Data adapted from K. Diem and C. Lentner, eds,
Scientific Tables, 7th edn, J. R. Geigy, Basle, 1970.

To release phosphorus from phosphate requires carbon, which in turn requires some form of organic matter to be decomposed by intense heat. Urine contains several organic chemicals that can act as a source of carbon, such as creatine, which is a constituent of all cells but especially of muscle fibre. Another abundant organic component is urea, but there are also others such as amino acids, various carbohydrates and enzymes.

Urine not only fascinated the alchemists but it also had important commercial uses in dyeing, scouring and the making of simple chemicals such as sal ammoniac (ammonium chloride) and saltpetre (potassium nitrate). The processing of urine had

---

in that way. The average adult passes around 120 g (4 ounces) of solid waste matter each morning, 75 per cent of which is water and 0.8 g of which is phosphorus. The total daily loss of phosphorus by the average adult is therefore 2.2 g.

started in the days of the Roman Empire, when vast quantities were collected for industrial use.

In *The Chemical Works of Caspar Neumann MD*, published in 1759 by the eponymous professor of chemistry at Berlin, there is a description of the components that could be separated from urine by distillation: spirit of urine, oil of urine and phosphorus of urine. The oil was even rumoured to taste as sweet as sugar. It was also possible, by slow evaporation, to grow crystals from urine which were known as microcosmic salt (sodium ammonium hydrogen phosphate, formula $Na(NH_4)HPO_4.4H_2O$). These play a part in the phosphorus story, as we shall see in Chapter 10.

According to Neumann, 60 gallons of urine could be evaporated to yield 10 pounds of dry matter which, on further heating, lost a third of its weight and from which eventually an ounce, or more, of phosphorus could be obtained. We learn from his book that, to get phosphorus, you needed a furnace of the type used by glass-makers. He also says that phosphorus could be purified by heating it in boiling water, stirring until the liquid phosphorus becomes clear. Phosphorus produced in this way was given a variety of names: *phosphorus urens*, *phosphorus glacialis* or *noctiluca glacialis*.

The volume of urine necessary to produce phosphorus using Brandt's method was truly enormous because his process was so inefficient. We know that Brandt told Leibnitz it would require 5 tuns of urine to produce a mere 2 loths of phosphorus.* The process was made unnecessarily complicated, as we can see from the first recipe published in English for making phosphorus. This is given in *Philosophical Experiments and Observations of the Late Eminent Dr Robert Hooke FRS*, which was written by a Dr Derham and published in 1726. It began as follows:

---

* A tun was a large wine cask capable of holding around 1,100 litres, while a loth was about 60 grams.

Take a quantity of urine (not less for one experiment than 50 or 60 pails full); let it lie steeping in one or more tubs till it putrify and breed worms, as it will do in 14 or 15 days. Then, in a large kettle, set some of it to boil on a strong fire and as it consumes and evaporates, pour in more and so on, till, at last the whole quantity be reduced to a paste and this may be done in two or three days if the fire be well tended, or else it may be doing a fortnight or more . . .

The account went on to describe how it was necessary to heat the residue in a retort for up to twenty-four hours in order to produce phosphorus. Derham also warned of its dangers:

Mr Concle [Kunckel] writ also with it on paper and the letters all shined in the dark . . . He once wrapped up a knob of it in wax at Hanover and it being in his pocket and he busy near the fire, the very heat did set it aflame and burned all his clothes and his fingers also; for though he rubbed them in the dirt, nothing would quench it, unless he had had water. He was ill for 15 days and the skin came off.

Brandt's method of making phosphorus, as described by Leibnitz, required several stages. The thick syrup that urine was reduced to was heated until a red oil distilled over and this was collected. The retort was allowed to cool and its contents consisted of a black and spongy residue in the upper part and a lower part that was salty. This salt was discarded. The black material was scraped off and mixed with the red oil, and the two were put back into the retort and heated for sixteen hours, during which time white fumes came off first, then another oil and then phosphorus, which was passed into cold water to solidify.

Brandt got only a little phosphorus because he did several things wrong. To begin with there was no need to leave the urine to putrefy. This does nothing to increase the amount of phosphorus, nor does it aid extraction. As others discovered later, using

The alchemy of phosphorus

Phosphate consists of a central phosphorus atom surrounded by four oxygen atoms and carrying three units of negative charge, in the form of electrons, so it is written PO_4^{3-}. These negative charges have to be counterbalanced by three positive charges, which might take the form of sodium (Na^+), this being the most abundant metal element in urine. To release the phosphorus from the strong grip of the four oxygens requires four carbon atoms to prise them off, thereby forming the gas carbon monoxide (CO). If sand is present, this reacts with any metals present to form silicates such as sodium silicate, Na_2SiO_3.

To generate these chemical reactions a great deal of heat is necessary. The overall chemical change is:

sodium phosphate + carbon + sand + heat → phosphorus + carbon monoxide + sodium silicate

Even today this is still essentially the same process for making phosphorus, except that calcium phosphate ore is used, the carbon comes from coke and the heating takes place inside an electric furnace.

fresh urine gave equally good results. Strong heating of the urine residue causes all kinds of chemical reactions to occur, driving off volatile organic matter as an oil, but this contains almost no phosphorus because phosphates are not volatile.

The residues left in the retort after this first heating can be identified as a black char of carbon, formed by the decomposition of organic material, and the salt layer. By discarding this Brandt was throwing away most of the phosphate, which explains why the amount of phosphorus he got was so small (less than 1 per cent of what would have been present in the urine). Had he ground the contents of his retort together to get an intimate mix of the carbonized layer and the salt and then heated that, he might well have increased his yield of phosphorus from 50 g to 500 g or even 5,000 g.

The method eventually published by Kunckel was similar except that he mixed the urine paste with fine sand before heating it in a retort. The use of sand gave an increased yield because it combines with the salts present, thereby aiding the decomposition of the phosphate.

Since all that is required to make phosphorus is urine and heat, it is reasonable to wonder whether phosphorus was, in fact, discovered prior to the seventeenth century. There are intriguing references throughout history to curious materials that might just have been phosphorus. For example, it might even have been known in Roman times. St Augustine, the Christian theologian and philosopher who lived from AD 354 to 430, wrote of 'perpetual' lights that were seen in the sepulchres of the early Christians. This is unlikely to have been a reference to phosphorus the element, but he might have been referring to another form of phosphorus, as we shall discover in Chapter 14.

The fourteenth-century French alchemist Achid Bechil referred to a curious 'carbuncle' that formed when he distilled urine mixed with clay, lime and other organic matter. Such a mixture might indeed give off phosphorus if heated strongly enough and this could explain Bechil's observations. But, because he does not refer to the carbuncle either glowing in the dark or bursting into flames, it seems unlikely that he had really made phosphorus.

More reliable was a report by the Swiss-born alchemist and physician Paracelsus.* He is best known as the founder of

---

* His real name was Theophrastus Bombastus von Hohenheim and he lived from 1493 to 1541.

modern medicine, and published a recipe in his *Archidoxorum* for the distillation of urine. In this report he says that water, air and earth will ascend together and the fire remains behind. All are recombined and distilled a second, third and fourth time, at which point the earth will remain behind. Cooling the material that distils produces 'icicles, which are the elements of fire'. This sounds tantalizingly like a reference to phosphorus. Yet, if Paracelsus really had discovered the real thing, he would surely have made more of his discovery and we are forced to conclude that his icicles were not phosphorus. In fact there is no convincing evidence to suggest there was an earlier discovery of phosphorus prior to the experiments of Brandt.

△

That phosphorus was an element was not appreciated at the time of its discovery. Another century was to pass before the French chemist Antoine Lavoisier first postulated that all matter is composed of *chemical* elements, and listed many of them, including phosphorus. Today we can place the discovery of phosphorus in the historical context of the discovery of the elements. It was the thirteenth chemical element to be isolated in its pure form.* Unlucky phosphorus.

To begin with, phosphorus was greeted with great acclaim, and yet it was damned from the moment it was born. It displayed properties that humans were in no position to cope with. As we shall read in Chapter 3, phosphorus promised cures but it delivered mainly curses. It is a deadly poison and yet soon after its discovery it was being sold by pharmacists as a treatment for all kinds of illnesses and especially mental conditions. Even more

---

* The others, in the order in which they were discovered, were: carbon, sulphur, copper, silver, gold, iron, tin, antimony, mercury, lead, arsenic and bismuth. These twelve occur naturally, or were easy to win from their ores, or were discovered by individuals unknown.

remarkable, it was to remain part of the medical pharmacopoeia well into the twentieth century, despite its having cured no one of anything in the previous 250 years.

While doctors used phosphorus, hoping to cure their patients, others used it to murder them (Chapter 9); and while some scientists were researching it with a view to making pesticides to benefit human beings, others were secretly turning it into nerve gases, the better to destroy them (Chapter 8).

Even Nature finds it difficult to control phosphorus, having assigned to it the role of limiting all life on Earth (Chapter 12). Phosphorus is in short supply, yet is essential for every living cell. However, when humans increase the amount in the environment by using it as fertilizers and detergents, the life-forms that flourish may not be the ones we want (Chapter 13).

Phosphorus has the power to burst into flames; again a mixed blessing. Its ability to burn was put to use in various ways down the ages, starting with phosphorus tapers and phosphorus matches (Chapter 4), and ending with phosphorus bullets and phosphorus bombs. The irony was that Hamburg was to be devastated by phosphorus in the twentieth century, when tens of thousands of its citizens would be burned alive by it (Chapter 7). Back in seventeenth-century Hamburg all this was well into the future, but, for good or evil, the genie of phosphorus had been loosed on the world.

Phosphorus was discovered when the practice of alchemy was giving way to chemistry. If a single chemical can be said to have precipitated that change, it was phosphorus. If a single event in the history of this element was responsible, it was Kraft's final demonstration of its remarkable properties at a private house in London one September evening in 1677, as we shall see in the next chapter.

2. The alchemist and his apprentice

In the winter of 1694, the latest edition of a scientific journal published by the Royal Society was on sale at the Prince's Arms bookshop in St Paul's churchyard. This particular edition was selling well because it contained a letter written fourteen years earlier but kept sealed until its author, Robert Boyle, died. It was rumoured to contain the secret of phosphorus manufacture upon which his one-time, impoverished apprentice had built a highly profitable business and made himself rich.

The Royal Society had been instituted by King Charles II in 1660 to promote the study of science, and its periodical went by the somewhat whimsical title of *Philosophical Transactions: Giving Some Account of the Present Undertaking, Studies and Labours of the Ingenious in Many Considerable Parts of the World*. There, in volume 17, page 583, appeared the scientific communication that was the talk of London:*

> **A paper of the Honourable Robert Boyle's, deposited with the Secretaries of the Royal Society, October 14 1680 and opened since his Death; being an Account of his making the Phosphorus, etc.**
>
> *Sept.* 30 1680. There was taken a *considerable quantity* of Man's Urine, (because the Liquor yields but a small proportion of the desired *Quintessence*) and of this a good part at

---

* I have modernized spellings to make this easier to read but retained the original emphasis and grammar.

least, had been for a pretty while digested before it was used. Then this Liquor was distilled with a moderate Heat, till the *Spiritous* and *Saline* parts were drawn off; after which the *Superfluous* Moisture also was abstracted (or evaporated away) till the remaining Substance was brought to the consistence of a somewhat thick *Syrup*, or thin *Extract*. This done, it was well incorporated with thrice its Weight of fine *White Sand*; and the Mixture being put into a strong Stone-*Retort*, to which a large *Receiver* (in good part filled with Water) was so joined, that the Nose of the Retort did almost touch the Water: Then the two Vessels being carefully luted together, a naked Fire was gradually administered for Five or Six Hours, that all that was either *Phlegmatic* or *Volatile* might come over first. When this was done, the Fire was increased and at length for Five or Six Hours made as strong and intense as the Furnace (which was not bad) was capable of giving: (which Violence of Fire is a Circumstance not to be omitted in the Operation.) By this means there came over good store of white Fumes, almost like those that appear in the Distillation of the Oil of *Vitriol*; and when those Fumes were passed and the *Receiver* grew clear, they were after a while succeeded by another sort that seemed in the *Receiver* to give a faint bluish Light, almost like that of little burning Matches dipped in Sulfur. And last of all, the Fire being very vehement, there passed over another Substance, that was judged more ponderous than the former because it fell through the Water to the bottom of the *Receiver*; whence being taken out, (and partly even whilst it stayed there), it appeared by several Effects and other *Phenomena*, to be such a kind of Substance as we desired and expected.

The kind of substance they desired and expected was phosphorus. The paper poses a question: why had its author, Robert Boyle, insisted that it not be published until his death? As we saw in the previous chapter, others on the Continent had published methods for making phosphorus, so there was no secret about it. If there is an answer to this question, it lies with the man who wrote it and his fascination with phosphorus. When the paper was deposited, Boyle had finally succeeded in making the element after more than two years of failed attempts. He was about to begin his ground-breaking researches into its properties, which we regard today as the very first experiments in chemistry. These required a regular supply of phosphorus, and it was the production of this, by his young assistant Ambrose Godfrey, which was to lead to London becoming the only place where the element was manufactured for the next seventy years. How this came to be so can be traced to events that occurred at Boyle's residence one evening in 1677.

The alchemist

Today Boyle is regarded as one of the founding fathers of chemistry. He was the brother of Lady Ranelagh and he lived at her London home, Ranelagh House, which was situated at 83/84 Pall Mall in the fashionable St James's district.* There, on Saturday 15 September, he and a group of fellow members of the Royal Society eagerly awaited the arrival of Daniel Kraft, the man who had recently come from Germany with samples of the new phosphorus.

Robert Boyle was a complex character: a lifelong bachelor, a staunch Christian, a giver to charity on a large scale, a scholar, a world-renowned scientist – and an alchemist. In 1677 he was fifty years old and, despite his ground-breaking work on the study of gases (leading to Boyle's Law, still taught today), he had spent a great deal of his life searching for the philosopher's stone. The fact that Boyle had been an alchemist for most of his life

---

* The Royal Automobile Club occupies the site today.

was to prove an embarrassment to the scientific establishment in years to come because of the need to present him as the first true chemist.

His famous book *The Sceptical Chymist* has been seen as the defining break between alchemy and chemistry, but Lawrence Principe's recent work on Boyle, entitled *The Aspiring Adept*, shows that this is not the all-out attack on alchemy that many imagine it to be. Sadly, when Boyle died his library was dispersed, but there is good reason to believe it contained many books on alchemy, and in fact Robert Hooke, the Secretary of the Royal Society, reported seeing these on sale in a London street market in March 1694.

The transmutation of base metals into gold was reported to have been achieved in the 1670s, and among Boyle's papers there was a *Dialogue on Transmutation and Melioration of Metals*, a work that was never published but the manuscript of which Principe has been able to reassemble from existing fragments. In it, Boyle describes one well-documented transformation of base metal into gold performed by a French alchemist and witnessed by several eminent people. Boyle believed his search for the philosopher's stone was justified because it would not only not transmute metals but would be an 'incomparable' medicine.

Boyle himself had written about transmutation in the Royal Society's *Philosophical Transactions*, where, on 21 February 1676, his paper 'On the Incalescence of Quicksilver with Gold' appeared. This reports on a 'mercury' that when mixed with gold causes it to react and evolve heat. 'Mercury' was the term used by alchemists to describe various materials and does not necessarily refer to the liquid metal element. Lord Brouncker, President of the Royal Society, attested to the efficacy of Boyle's new 'mercury' in that, when it was mixed with gold powder on the palm of his hand, he felt the heat it generated.

In another work, *Producibleness of Chymical Principles*, Boyle reports on a 'mercury' that could dissolve gold instantly but refuses to reveal its nature because he feels it would 'disorder the affairs of mankind, favour tyranny and bring a general confusion, turning the world topsy-turvy'. We can only guess what this 'mercury' was, but Principe says that by following the instructions of an American alchemist, George Starkey (a contemporary of Boyle), he was able to make an antimony–mercury compound that did give off heat when mixed with gold. Boyle's instructions for making it, though, were written in alchemical code: 'Take pure Negerus, Dakilla, imbrionated banasis ana, mix them very well together & drive off all that you can in a retort with a strong fire of sand. It dissolves gold readily and that with sensible heat.'

Negerus was mercury, imbrionated banasis was antimony, Dakilla was copper. The danger inherent in carrying out such an experiment is mercury-vapour poisoning and this might well explain Boyle's chronic sickness, as it did that of his fellow alchemist, Isaac Newton.* Mercury may have hastened Boyle's death and even that of his sovereign, King Charles II, who also dabbled in alchemy and had a resident 'chymist' at Court, Nicholas le Fevre. Charles officially died of 'apoplexy', although it has been suggested by Professor Frederick Holmes of the University of Kansas Medical Center that acute mercury poisoning was really the cause. Holmes has re-examined accounts of the King's sudden illness, which started on Monday morning, 2 February 1685, terminating in his sudden death four days later. An autopsy was performed on Saturday the 7th and Holmes says its findings are consistent with acute mercury poisoning.† Days

---

* Isaac Newton, who also dabbled in alchemy, was affected by mercury vapour around this time although he lived well into the next century, dying in 1727. An analysis of a strand of Newton's hair showed abnormally high levels of mercury.

† The best medical treatment then available was used in an effort to save the King. He was bled, given emetics, purged, blistered and prescribed extracts of human skull, but to no avail.

before his final illness the King had been heating mercury in his laboratory at the palace and may well have inhaled its highly toxic vapour.

Undoubtedly Boyle was exposed to mercury fumes as well, but there is no evidence that he was disabled by them. He had taken up residence at Ranelagh House in 1671 and lived there until his death in 1691. In 1676 he persuaded his sister to allow him to build a laboratory in the garden and then to enlarge it in 1677. It was equipped with a furnace, retorts, flasks and other alchemical apparatus together with a range of simple chemicals with not-so-simple names like oil of vitriol (sulphuric acid), lunar caustic (silver nitrate), sugar of lead (lead acetate) and butter of antimony (antimony chloride).

Little of what he did there in pursuing his alchemical interest was worth reporting, although he may have spoken about his experiments when he attended the Royal Society. He may even have been investigating phosphorescent materials since we know there had been much talk of these at recent meetings. Boyle had given his opinion that this was a subject worthy of study because God in his infinite wisdom had set aside the fifth day of Creation for the making of light. The previous March, Boyle had given a talk about various phosphoruses and he had written about the luminescent Bologna stone and so-called Baldwin's phosphorus in issues of the *Philosophical Transactions of the Royal Society*.* Then they learned of a much brighter phosphorus from a letter sent from Germany to Robert Hooke, the Secretary of the Society, which described Kunckel's phosphorus, and at their meeting on 17 May Hooke reported on a communication he had received from Leibnitz describing the phosphorus he too had

---

* In 1676 Christian Baldwin had sent a specimen of his 'phosphorus', which was a form of calcium nitrate, to the Royal Society where, at a meeting of its members on 4 January 1677, it was examined and its luminescence shown to be genuine. Baldwin was duly elected a fellow.

made. Hearing about the new phosphorus was one thing, seeing it was another, and it was with great pleasure that they learned that the famous Dr Kraft had been invited by King Charles to come to London. The Court were eager to see it and were willing to pay his fee of 1,000 thalers.* This seemed a trifle high, but as he was the only person able to put on a display of this new wonder, it was deemed a necessary expense if the royal family were to witness the remarkable spectacle that so many of their continental cousins had already seen.

When Kraft arrived in London, Boyle contacted him and invited him also to put on a display of the new phosphorus for the benefit of the fellows of the Royal Society. What transpired at Ranelagh House on that historic night marked the turning point from alchemy to chemistry – at least for Boyle. What he saw impressed him so greatly that he became obsessed with phosphorus, and this obsession was to lead not only to his finding a way to make it for himself, but to some real and remarkable chemical research into its properties.

The demonstrations of Daniel Kraft

Shortly after supper on that Saturday evening in September, as it was growing dark, the arrival of Mr Daniel Kraft at Ranelagh House was announced. Boyle and the other fellows of the Society greeted him warmly. He had brought with him a large box and began carefully to lay out its contents on the table. There were several tubes of different lengths and glass vials containing various liquids, but his most impressive piece was a large hollow sphere, about five inches across (12 cm), which contained two spoonfuls of a reddish watery material and whose tiny opening was closed with sealing wax. However, Kraft took particular care

---

*   The equivalent today would be around £20,000 or $30,000.

with a small bottle containing a lump about the size of an almond. When his display was ready, he asked that the window shutters be closed and all the candles removed: the show was about to begin.

First the glass sphere was passed around and, according to Boyle, this shone 'like a cannon bullet taken red hot out of the fire, except that ... it was more pale and faint. But when I ... shook it a little, the contained liquor appeared to shine more vividly and sometimes to flash.' The fellows next examined a glass tube in which at one end were tiny pieces of phosphorus that seemed to make the whole tube glow. Kraft then took one of the vials containing a little liquid and shook it, whereupon there was a 'flash of lightning' within and smoke filled the vessel.

The time had come to move on to Kraft's *pièce de résistance*, the bottle with the lump of phosphorus. This was passed around while Kraft explained that the lump inside had glowed continuously for two years. Boyle observed, as he held it in his hand, that there was no sign of smoke or fumes yet the material seemed to glow even more vividly than the liquids they had already been shown. Kraft took some of this phosphorus and broke it into tiny fragments which he then scattered on the carpet, 'where it was very delightful to see how vividly they shined'. Boyle, worried that they were damaging Lady Ranelagh's Turkish carpet, called for candles to be brought into the room but the carpet was found to be unharmed.

The candles were taken away again and the demonstration continued. Kraft now took a piece of paper and, having wiped some of the phosphorus on to the tip of his finger, he wrote the word DOMINI on the sheet in glowing letters, which Boyle said was a 'mixture of strangeness, beauty and frightfulness'. As he examined the paper, he was aware of a strange smell resembling sulphur and onions. When Kraft put the glowing tip of his finger to his face, it gave off enough light to see his features. He rubbed the piece of phosphorus on the back of Boyle's hand and the cuff of his jacket, both of which now shone 'very vividly'. When he tried to wipe the phosphorus from his hand it was not to be moved and continued to shine but without any noticeable heat being generated.

To show the spontaneously flammable ability of phosphorus, Kraft tried to ignite a small pile of gunpowder with it. He scooped a tiny piece of phosphorus on to the end of a quill pen and placed it on the gunpowder, but nothing happened. He tried another piece but still it did not work. Boyle suggested that the powder was to blame, suspecting that it might be of inferior quality or slightly damp. Somewhat embarrassed by his failure to ignite gunpowder, which was clearly the climax of his performance, Kraft told Boyle he would come back the following weekend and try again. Nevertheless, what they had seen was more than enough to impress the group of fellows and they congratulated Kraft. But before they could ask him about the new phosphorus he begged leave to depart since the hour was late.

He returned to Ranelagh House the following Saturday afternoon and this time he had brought his main supply of phosphorus. This he kept under water and removed it by sticking the end of a quill pen into it and pulling it out of the bottle; he then dried it with blotting paper, whereupon it began to smoulder. Kraft quickly cut off a piece about the size of a pin head and dropped the bulk of his phosphorus back into the water. He cut the tiny piece of phosphorus in two, and one of these he spread on a clean sheet of paper which he then warmed until it flashed and burned where the phosphorus covered it. The other piece of phosphorus he 'put upon the tip of a quill and having ... very well dried and warmed some gunpowder upon another piece of paper, he laid that paper upon the ground and then holding his quill upon it ... within half a minute (by my guess) that powder took fire and blew up'.

Impressed by this, Boyle invited Kraft to stay a while and

they discussed the nature of phosphorus. Boyle speculated on its source of light, suggesting that, if this depended upon the presence of air, then by placing it inside a glass vessel and pumping the air out, the light would go out. Would Kraft leave some of his phosphorus with Boyle so he could carry out this experiment – or at least say how phosphorus was made? Kraft declined to do either.

Boyle was clearly disappointed and offered to exchange a secret piece of alchemical information, which he claimed he had told no one, in return for the recipe. Kraft was intrigued and Boyle told him about the new 'mercury' he had made, but Kraft feigned a lack of interest, although he was prepared to give Boyle a clue to the source of the new phosphorus. 'It was', he said, 'somewhat that belonged to the body of man.' And that's all he would say.

Once Kraft had departed Boyle pondered what this meant and decided the 'somewhat' must be urine. He had already used this in some of his own alchemical experiments and had a quantity of it stored in his laboratory, but it was to be almost a year before he could begin experimenting with it. More pressing matters had to be dealt with and we know from a letter he wrote to a Dr J.B. (probably Johann Becher) that he was distracted 'by divers removes, indispositions of the body, law-suits and other avocations'.

The apprentice

Some time in the year following Kraft's visit to London, Boyle turned his mind to the making of phosphorus. He had in his employ a man by the name of Bilger, of whom little is known, and it fell to him to boil down large volumes of urine and reduce it to a thick paste. But how to extract phosphorus from it? Boyle consulted his books on alchemy to see what they had to say about urine but, no matter what he and Bilger tried, no phosphorus was forthcoming.

Thinking Kraft might also have been alluding to faeces, he even got poor Bilger to heat the contents of the Ranelagh House night-soil pits, but still he could not make it. In 1678 or early 1679, Boyle took into his employ the alchemist Johann Becher, who had come to London seeking work and who had previously been alchemist to the Duke of Mecklenburg-Güstrow. On his advice, Boyle also recruited seventeen-year-old Ambrose Godfrey Hanckwitz to help in his laboratory. The young Godfrey, as he preferred to be called, came from Nienburg in Saxony and had been Becher's apprentice in Germany. Neither Becher nor Godfrey knew how to extract phosphorus from urine but they knew a man who did: Brandt. When Godfrey next paid a visit to Germany he went to Hamburg and called on the alchemist, who gave him the vital piece of information they needed: very high temperatures had to be used to drive off phosphorus.

On his return, Godfrey immediately set to work on a new batch of urine using all that Boyle had collected from the privies of Ranelagh House. He did as Brandt had told him and heated the urine residues until they were red hot – so hot that the retort cracked. A disappointed Godfrey went to tell Boyle what had happened and Boyle returned with him to the laboratory to see the damage for himself. Before the retort had broken, a vapour had distilled from it, said Godfrey, and had condensed as a white solid in the receiver to which the retort was connected but it was not phosphorus because it was not shining. Boyle took the flask into a dark corner ('though it was then day') and there, to his delight, he could see the faint glow of phosphorus.

Boyle's first published account of phosphorus was written in 1680 in the form of an open letter to Dr J.B. in which he spoke of the new phosphorescent material he called *aerial nocticula* (spirit of night light). In it, he speculated about the uses of phosphorus and thought it might provide safety lights for the holds of ships. Several warships were known to have been destroyed when sailors with burning flares went below decks to get gunpowder. He also thought that phosphorus, sealed in glass bulbs, might be lowered into the sea to attract fish. He knew that in Scotland and Ireland some fishermen used candles in jars for this purpose. He even suggested that phosphorus could be used to illuminate the dial of a clock so that you could tell the time if you woke in the middle of the night: '. . . but these trifles, though pretty, I purposely pass over; as also a use that may be of great but I fear of mischievous consequence'. What that use was we can only guess, but he went on to speculate that 'the utilities of so subtle and noble a substance may be brought to afford in medicine, may be more considerable than any of its other particular uses'. Like others in Germany, Boyle believed that something so remarkable as phosphorus must have healing powers, although he never reported trying it himself. Nor did he treat it as the long-sought-for philosopher's stone. His interest in the new material was, by now, purely scientific.

The chemist and his laboratory assistant

Looking through the papers that Boyle wrote on phosphorus, we can see that by 1680 he was no longer a secretive alchemist but a scientific chemist. He carried out experiments, recorded his observations, tried to explain what was happening and published his results in plain English. But, more than that, he wrote them up in a way that made them accessible to ordinary people. It is not difficult to see why Boyle is today regarded as the founder of modern chemistry – and why generations of scientists have been keen to deny his alchemical past.

As Boyle gave up alchemy for chemistry, so Godfrey turned from apprentice to laboratory assistant. His job was now to make phosphorus for Boyle on a regular basis and gradually he perfected the method of doing so. Boyle had the new phosphorus in two forms: a liquid and a solid. It is clear from his descriptions that both were phosphorus and differed only in their degree of purity. Pure phosphorus melts at the relatively low temperature of 44°C (111°F), which is only a few degrees above body heat. It is very easy to depress the melting point with impurities and the liquid form was phosphorus mixed with oil of urine.

Boyle said of liquid phosphorus that it did not shine while in a well-stoppered bottle but that, when he loosened the cork and air was admitted, he could see tiny flames and the contents began to glow; this would last for as long as three hours after he had replaced the cork. He also observed that air rushed in when he removed the cork from the vessel. He attributed this to 'debilitation' of the air inside the flask, but we now know that it occurred because some of the phosphorus had reacted with the oxygen in the air, thereby creating a partial vacuum. (Oxygen was not to be discovered for another hundred years.)

Boyle noted that when he rubbed the liquid phosphorus on his hand it would shine vividly and sometimes would flame and flash and give off an offensive smoke. The effect would gradually die away but was revived by further rubbing, although eventually the effect disappeared altogether. He tasted the liquor and found it was neither acid nor alkaline.

On another occasion he took a bottle of phosphorus up to his bedroom and, waking before dawn, put it under the bed-clothes and noted that it shone 'prettily' and that while he could easily see his fingers he could not make out the colours of the red ruby and green emeralds set in a ring that he wore. One night he slept with the vial of phosphorus next to him in bed and was amazed in the morning that it was shining more brightly than ever, which he attributed to its being warm.

His favourite trick was to dip his finger in the liquid phosphorus and draw lines on linen or on the hands of guests, both men and women, noting how a little phosphorus went a long way and that those hands he had touched could then touch others and pass the luminosity on. When he dipped a paint brush into the liquid phosphorus it appeared as if it were aflame, burning like a wax taper. Boyle speculated on the nature of the phenomenon he was witnessing and compared it with other inexplicable forms of light such as fiery meteors, shooting stars, summer lightning and Will-o'-the-wisp. (His views on Will-o'-the-wisp were nearer the mark than he could have imagined: see Chapter 14.)

Godfrey also produced solid phosphorus for Boyle. On one occasion, when he used a knife to cut it, Godfrey wiped the blade on his calico apron, using the thumb and index finger of his left hand. Immediately he felt a sharp stab of heat and found two holes burned in the apron where it had been pressed against the knife. Boyle inspected the damage and concluded, rightly, that the solid form of phosphorus was more potent.

Boyle observed that a mixture of phosphorus and sulphur ignited when the two elements were mixed and crushed together. They did not burn away with a slow flame but blazed up almost like fired gunpowder, except that the flame was more luminous. In another experiment he burned an ounce of phosphorus in a bell-jar and obtained a white deposit on the glass. When this was scraped off and left in the open, it quickly picked up water enough to cause it all to dissolve and produce an acid as strong as sulphuric acid.* Godfrey drew his attention to a red deposit in the bell-jar which, on heating, burst into flames, and this would appear to have been the first example of red phosphorus, the safer form of the element, but the observation was not followed up and this form of phosphorus was finally rediscovered 160 years later.

Why, then, in the autumn of 1680 did Boyle deposit the

---

* Phosphorus burns to form phosphorus pentoxide, which absorbs water vapour from the air to form phosphoric acid.

sealed envelope with the Secretary of the Royal Society, with instructions that it was not to be opened until his death? I suspect it was to ensure that the secret he had uncovered would one day be made public knowledge. Maybe he regretted his action because at the end of one of his papers about phosphorus he reproduced the contents of the sealed letter. Some tried to follow these instructions but usually failed to make any phosphorus, and they assumed he had deliberately omitted some vital step in the process which would be revealed one day when the letter was opened. Those who waited a further fourteen years, hoping to discover what this secret was, were to be disappointed because the sealed letter contained nothing new.

In 1681, Boyle published more information about phosphorus in a booklet entitled *New Experiments and Observations made upon the Icy Noctiluca*. Boyle comments on the drought of that year: 'a season whose dryness continued almost to a wonder', which forced him to leave London and take refuge in a small village outside the capital to escape the worst of the heatwave. He took his samples of phosphorus with him and continued his investigations, measuring how long phosphorus would shine for and under what conditions. One evening, he entertained a lady with a demonstration during which he swirled a glass vessel until it was fully coated with liquid phosphorus and then removed the stopper. The result was dramatic, with the vessel not only shining but 'twinkling . . . like so many little stars in a cloudless but dark night and continued this scintillation longer than one would have expected . . .'

Godfrey stayed behind in the capital making more phosphorus and getting more proficient at doing so, obtaining lumps of it 'sometimes as large as small beans and sometimes at least three or four times as large'. These were colourless and transparent when held against the light, which is why Boyle named it icy or glacial noctiluca, and it is obvious to modern chemists that they were producing pure material. Sometimes the phosphorus was reddish in colour, sometimes a 'pleasing blue' and 'sometimes of a colour to which I cannot easily assign a known name'. Boyle observed that, when he dipped a piece of phosphorus in water, the portion below the surface ceased to shine while that held in his fingers continued to do so.

Boyle also investigated the water under which phosphorus had been stored, noting that it had a strong and penetrating taste not unlike that of sea-salt and vitriol (sulphuric acid) together. Slow oxidation of the phosphorus would produce phosphorus oxides which would dissolve in water to give solutions of various acids. When he boiled off the water, he was left with a semi-solid substance which we can recognize as phosphoric acid. If this was heated strongly, it melted and could then be drawn into threads of 'perhaps a foot or more in length', which demonstrates that he had produced a mixture of polyphosphoric acids.

When the solid produced by burning phosphorus was left all night it picked up moisture and became liquid and, when this in turn was boiled again to dryness, it crackled with flashes and bangs and 'what seemed strange, that though oftentimes two and sometimes more flashes appeared at once, yet so small a quantity of matter continued to afford them for almost an hour together but that the late time of the night obliged me to go to bed before the experiment was finished'.

He tried dissolving phosphorus in other liquids, such as oils and spirits of wine (alcohol). He even put a piece in concentrated oil of vitriol (concentrated sulphuric acid) and noted no reaction. He found it would dissolve in some oils, such as that of cinnamon, and this also disguised its smell, which 'the delicate sort of spectators and especially the ladies' found offensive. By this means, he could even demonstrate its luminosity to 'persons of quality'. But oil of cinnamon was expensive and more often he used oil of cloves, which was cheaper.

The increasing solubility of phosphorus went in the order: water, alcohol and the various oils. The nearest he came to actual measurement was his observation that a grain of phosphorus (which is about 65 milligrams) would 'impregnate' four or five ounces of alcohol (which is about 100 grams). With some of these solutions he could produce the luminosity he desired. For example, pouring the alcoholic solution into hot water produced more light than when it was poured into cold water. He tried diluting the solution of phosphorus in alcohol with water and noted that the luminosity was still observable even though he diluted it to a hundred-thousandth of its original concentration. He repeated the experiment again and diluted it to four hundred-thousandths of its original concentration and still it was luminous. (The ability of a tiny amount of phosphorus to reveal its presence in this way enabled forensic scientists in the twentieth century to detect tiny amounts in murder victims, as we shall see in Chapter 9.) Boyle reported on the dense smoke produced when phosphorus burned, again a property the twentieth century was to make use of to provide smokescreens in warfare – Chapter 7.

These experiments were not without their hazards, as Godfrey was to discover as he helped Boyle. When he squeezed a piece of phosphorus between his fingers it felt hot and scorched the skin, but this did not deter him from handling it with his fingers, so they were often covered in blisters which were extremely painful and slow to heal. On another occasion, Godfrey was asked to demonstrate how phosphorus could ignite gunpowder but nothing seemed to happen when he put a piece of phosphorus on a pile of powder. As he leaned over to see what was wrong, it suddenly burst into flames and set fire to his hair. As Boyle observed: 'This, proving innocent enough, became more diverting than the smell of smoke that succeeded it.'

One day Godfrey was taking a vial of phosphorus to Boyle in the village, carrying it in the pocket of his trousers, when it broke. 'Whereupon the heat of his body, increased by the motion of his long walk ... did so excite the matter, that was fallen out of the broken vial, that it burned two or three great holes in his breeches, the recent effects of which I could not look upon without some wonder as well as smiles,' wrote Boyle.

Two years of research finally exhausted Boyle's curiosity in phosphorus and so he wrote two accounts of his work. The first, entitled *The Aerial Noctiluca, or some new phenomena and a Process of a Factitious Self-shining substance*, was published in 1681 and this provided a general description of phosphorus. The second, *Observations made upon the Icy Noctiluca, imparted to a Friend in the Country, to which is annexed a Chymical Paradox*, was published in 1682. The observations in this latter tract included reports on solvents for dissolving phosphorus and the element's reaction with various acids and oils; on the effect of leaving phosphorus exposed to the air; on the flammability of phosphorus; on the reaction of phosphorus acids and metals; and many other experiments. Considering the limited resources at his disposal, the results he obtained were truly remarkable and no real progress was to be made in the chemistry of the element for another century.

The spin-off company

Godfrey's interest in phosphorus also continued, but by 1682 he too was keen to move on. His years with Boyle had been very instructive but they had also been years of bitter argument with Becher.

Godfrey left behind a large collection of personal papers which survived into the nineteenth century, when they were examined by Joseph Ince, a writer for the *Pharmaceutical Journal*. Among these papers Ince found a short tract entitled 'An Apology and Letter touching a Crosey-Crucian' which appears to be Godfrey's allusion to Becher, who was a Rosicrucian.* Apart

---

* The Rosicrucians were a secret society whose symbol was a rose and a

from a few extracts quoted by Ince, nothing survives of this or any of Godfrey's papers, but we can piece together some of what happened.

When Godfrey arrived in London in 1679 he rented rooms in Chandos Street in the Covent Garden area and lived there with his young wife. Becher shared the same lodgings and, to begin with, the two of them got on well together, especially as Becher was able to minister to Godfrey's wife when she was sick. It appears that the alchemist performed his experiments at this address but became depressed when they did not work and said he was missing his wife and family. Godfrey reported this to Boyle, who paid to have Becher's wife and daughter brought to London, where they joined him in Chandos Street.

Things deteriorated soon after Becher's wife arrived and she seems to have taken an instant dislike to the young apprentice. She regularly made scenes outside Godfrey's rooms, accusing him of all kinds of misdemeanours. With Godfrey's help, the Becher family moved to other lodgings but things continued to go from bad to worse and Becher's wife kept returning to Chandos Street, banging on Godfrey's door and shouting abuse in German, demanding money that she claimed he owed them, suggesting that Godfrey was unfaithful to his wife and, worst of all as far as Godfrey was concerned, accusing him of cheating Boyle.

The disputes between Godfrey and Becher came to a head when Boyle reduced the alchemist's salary, as nothing appeared to be forthcoming from his experiments. Becher blamed Godfrey

---

cross, the Latin names of which gave the brotherhood its name, although its adepts believed it was founded in 1484 by Christian Rosenkreutz and took its name from him. He is now regarded as a mythical figure, and the society was probably started by Paracelsus in the following century. The Rosicrucians claimed to be the depository of ancient wisdoms, and the cult was particularly attractive to alchemists.

for this cut in his income and he went with his wife to Chandos Street, where an almighty row ensued, with Becher's wife now saying that the Godfreys had ruined them. She became convinced that Godfrey's wife was the cause of her troubles and took to following her in the street, shouting and spitting but, 'thank God, all in German, that the people understood her not,' wrote Godfrey.

Meanwhile Becher went to see Boyle, but he had heard of the quarrels with Godfrey and refused to see him. Becher then returned to Chandos Street for another row with Godfrey and that is the last we hear of him. He appears to have died that same year. In his letter, Godfrey muses how unwise it is 'meddling with Rosicrucians and mad-men'.

Free of the harassment of Becher and with some financial help from Boyle, Godfrey began to manufacture phosphorus. In 1683, according to the records of the nearby church of St Martin-in-the-Fields, the Godfreys' first son was born, and he was named Boyle Godfrey in honour of the great man.* By 1685 his father had built up a flourishing business, advertising solid phosphorus for sale at 50 shillings wholesale and £3 retail per ounce. Godfrey had a furnace at the rear of his lodgings where he made the material and he had found a good way of purifying it, which consisted of melting it and squeezing it through chamois leather. Not surprisingly, on one occasion his hands got badly burned and for three days he lay in great pain. Despite such setbacks, Godfrey slowly began to build up a successful enterprise and we will take up his story in Chapter 10.

△

---

* He and his wife had two other sons, Ambrose and John. Boyle's wife died in the early 1700s and he married his second wife, Mary, on 15 October 1706. They had no children and in the end she outlived him by twelve years, dying in September 1753.

The obsession of the fellows of the Royal Society with phosphorescent materials was satirized in the Restoration comedy *The Virtuoso*, written in 1776 by Thomas Shadwell, in which Sir Nicholas Gimcrack claims to be able to read his Bible by the light of a leg of pork. The fellows were not amused and saw the play as yet another attack on them. These attacks were motivated by classical scholars of the day who doubted that wisdom could be achieved by mere observation and experiment. The fellows' interest in phosphorescent materials laid them open to mockery and the charge that such effects were little better than stage tricks. The production of phosphorus for sale, by Godfrey, was to be instrumental in reversing this opinion.

The diarist John Evelyn went to a phosphorus demonstration given by Frederick Slare at Samuel Pepys' house in 1685 and was greatly impressed: 'This phosphorus was made out of human blood and urine, elucidating the vital flame or heat, in animal bodies. A very noble experiment.' The sample of phosphorus used was almost certainly produced by Godfrey, although not from blood, and he continued to supply Slare, who staged demonstrations of phosphorus for many years, one of his favourite tricks being to pass samples of phosphorus dissolved in oil of cinnamon among his audience, inviting people to daub it on their hands and faces.

Godfrey found he could easily sell all the phosphorus that he made and he even began to export it to mainland Europe. All kinds of people were interested in buying it: scientists who were interested in its curious properties; showmen who could use it to delight audiences; and doctors who were starting to prescribe it for various ailments. This last group provided a growing outlet for Godfrey's phosphorus as people began to take it as a medicament that was reputed to cure almost any ailment, despite the fact that it is one of the most toxic of the chemical elements – and never cured anyone of anything.

3. The toxic tonic

Fabre reports that ointments of phosphorus have been known to inflame the skin and the wounds are deep and serious. He advises that no liniment or ointment should contain a greater proportion of phosphorus than one grain to the ounce. As to the possibility of its absorption by this channel in an active condition, De Lens affirms that he has been told by those who were in the habit of handling phosphorus frequently that they suffered from venereal excitation . . .

These warnings are taken from Dr Ashburton Thompson's book *Free Phosphorus in Medicine*, which was published in 1874 and which was devoted to extolling the benefits this element could bring in the relief of human suffering. Phosphorus ointment itself was never very popular as a cure for skin complaints, although it might have fascinated users as it was reputed to cause them to glow in the dark. More fascinating was the possibility of phosphorus producing 'venereal excitation', which was Thompson's polite euphemism for sexual arousal. Almost from the time it was first used in medicine, and possibly because of the source from which it originally came, people believed that it was an aphrodisiac.

By the time Thompson's book was published, phosphorus had been known for about 200 years and, during most of that time, had been used in medicines. Today it seems odd that a deadly poisonous material could be employed as a healing drug but, in the context of the time when it was discovered, this was no bar to its use. Indeed, all manner of dangerous chemicals had been advocated as possible cures by no less an authority than the great but controversial Paracelsus.

Paracelsus was born at Einseideln near Zurich in 1493. (He died at Salzburg in 1541 in mysterious circumstances.) Paracelsus opposed the ancient idea that illness was due to an imbalance of 'humours' within the body. Instead he taught a revolutionary doctrine, known as iatrochemistry, that insists each disease has a specific cause and therefore a specific *chemical* remedy. It was thanks to this novel idea that all sorts of chemicals were tried as medicaments, for example compounds of mercury for treating syphilis, those of antimony for treating fevers and those of bismuth for stomach upsets. The fact that mercury and antimony are highly poisonous did not prevent them being prescribed.

It was to be expected, given such an approach to healing, that equally deadly phosphorus would find a place in the medical pharmacopoeia. Indeed Boyle had speculated along such lines. The fact that it was extracted from urine and glowed with its own source of light only added to its attraction, and this glow was taken as strong evidence that phosphorus really was the 'flammula vitae', the vital flame of life.

The first person to sell phosphorus as a medical treatment was Johann Lincke, the German apothecary who ran the Golden Lion pharmacy in Leipzig. There was a slight problem, however, because phosphorus was flammable and so it was necessary to give it a protective layer. Lincke found he could do this with a coating of silver or gold and simply immersed the phosphorus in solutions of either silver (probably the nitrate) or gold (probably the chloride). The reaction of phosphorus with the dissolved metal formed a layer of silver or gold on the surface of the pill.

While these metals might have provided sufficient protection to prevent a pill from rapidly oxidizing and thereby catching fire, they were incapable of preventing its slow oxidation, which happily converted most of the deadly phosphorus to harmless oxides and acids.

Lincke started to sell phosphorus pills as 'Kunckel's pills', using the late professor's name to boost their appeal. (Kunckel died in 1702.) Each pill weighed three grains (200 mg) and they were reputed to be of benefit in treating colic, asthmatic fevers, tetanus, apoplexy and gout. They were, of course, useless in treating any of these conditions, and if they really had contained 200 mg phosphorus they would have caused severe poisoning. There were instances of much larger pills, with up to 12 grains (800 mg) of phosphorus, being prescribed which would undoubtedly have killed the patient had they been pure.

Kunckel's pills ceased to be manufactured when Lincke died. He never revealed his method of making them, and so for a hundred years they were unavailable until a French pharmacist, Alphonse Leroy, rediscovered the means of their manufacture in 1798. He soon realized that a single pill could be fatal and he too refused to reveal his method for making them and so again they disappeared from the pharmacopoeias of Europe, this time never to return.

Phosphorus was known to doctors in the early 1700s either as *phosphorus fulgens* or as *phosphorus anglicanus*, the latter name reflecting the country of origin from which most of it came. It could be prescribed as a tincture (a solution in alcohol), dissolved in an edible oil or rubbed on the skin as liniment. Whatever the preparation, the phosphorus absorbed by the patient was likely to be mostly oxidized and thereby rendered harmless. Despite its danger – or ineffectuality – there were soon glowing reports of phosphorus curing a variety of diseases and relieving all manner of conditions. The first written account of it as a medicament was in *Bates Dispensary*, published in 1720.

Table 3.1: Phosphorus in the various parts of the human body
(The total phosphorus in the average adult, weighing 70 kilograms, is 840 grams)

Body part	Amount of phosphorus (in grams) per kilogram fresh weight
Teeth	130
Bone	110
Brain*	3.4
Liver	2.7
Muscle	1.8
Kidneys	1.8
Lungs	1.6
Heart	1.5
Skin	0.4
Hair and nails	nil
Whole body average	*12*

* Average for grey and white matter. Grey matter is 4.0 g, white matter is 2.2 g.

Phosphorus on the brain

What really gave phosphorus treatments a boost though was a discovery made by Professor Johann Thomas Hensing (1683–1726). In 1719 he detected phosphorus in brain tissue and concluded that the amount to be found there was unexpectedly high. He was right: the human brain contains larger amounts of phosphorus than any other organ of the body. Total phosphorus in grey matter is 4 g per kilogram (fresh weight), whereas for the liver it is 2.7 g and for other organs is much less, as Table 3.1 shows.

The implication of Hensing's finding was that phosphorus was particularly important to the brain. Indeed it seemed right that so miraculous a material should have a role to play in the most vital part of the body. This belief was to persist in one form or another until recent times. The link between phosphorus and the brain was further strengthened in 1730 when a Dr Kramer announced that phosphorus was a remedy for epilepsy and melancholia (depression).

Hensing's research provided an explanation as to why humans differ so much in terms of mental ability: too little phosphorus and we lack intelligence and are dull and stupid; too much phosphorus and we are in danger of mental instability. In terms of his theory the more phosphorus our brain contained, the brighter we should be. The theory also explained the odd behaviour of the highly intelligent and why those who were clearly geniuses so often displayed an eccentricity seeming to border on madness. What the normal person should aim for was just the right amount of phosphorus. Of course the theory was nonsense, but that did not stop it being widely believed.

In the nineteenth century the philosopher Friedrich Nietzsche (1844–1900) carried this theory one stage further. He thought that chemicals in the body might have a determining influence on personality and sought to explain various character traits this way. For example, he thought bilious people lacked sodium while those with a melancholic outlook lacked potassium, and the phlegmatic type of person he claimed was short of calcium phosphate. On the other hand, a courageous few could attribute their bravery to an excess of iron phosphate. Like Hensing's ideas of the preceding century, those of Nietzsche were equally fallacious.*

Phosphorus acquired an undeserved reputation as a treatment for TB as the result of work by Leroy, in 1796. He had seen a patient who was on her deathbed from the disease and who

---

* Nietzsche was certified insane in 1889 when he was forty-five years old.

pleaded for help. Unable to prescribe anything that he knew would work, he decided to experiment with water from a jar in which some phosphorus was stored. He dissolved sugar in it and gave the solution to her to drink. To his surprise she seemed to improve and, indeed, she managed to survive a further fifteen days against all the odds. He gave the same medicine to other TB patients and they too appeared to respond positively. Impressed by the effect that his phosphorus water was having, he prescribed it for all kinds of ailments, generally with a beneficial outcome.

Convinced of its curative effects, Leroy even took a phosphorus pill and recorded the effect it had on him. On the first day he experienced a feeling of nausea, thirst and a heightened sense of excitement. On the second day, he felt unusually fit and well and this continued for several days, accompanied by improved digestion. As we shall see, he was describing the symptoms of poisoning by a less-than-lethal dose on the first day, and the effects associated with recovery in the days thereafter. Nevertheless, Leroy felt sufficiently confident in the therapeutic value of elemental phosphorus that in 1798 he published a series of articles advocating its use in the treatment of many serious illnesses from consumption to cholera. He even tested its aphrodisiac qualities on listless ducks and apparently they responded extremely well.

Inevitably, therefore, it became popular to drink the water in which a piece of phosphorus was stored. The water could be replaced regularly as it was used up and a stick of phosphorus would last for years. The solubility of phosphorus in water is only 2.3 mg per litre and much of the phosphorus that dissolved would be oxidized by the air and hydrolysed by the water, rendering the 'tonic' relatively safe.

Those who wanted a more potent dose of phosphorus than phosphorus water could deliver took to dissolving the element in organic solvents such as almond oil or alcohol, in which it is far more soluble. Drinks like rum and brandy could even take up enough phosphorus to deliver a fatal dose, as we shall see in Chapter 9.

Clearly taking phosphorus water was much safer than drinking such tinctures or even taking phosphorus pills, but, as doctors became aware of the dangers inherent in prescribing the element, the recommended medicinal dose of phosphorus decreased, until by the mid-nineteenth century the standard dose was only one-eighth of a grain (8 mg) and by 1900 it was more normal to give one-twentieth of a grain (3 mg) or even as little as one-hundredth (0.7 mg). Preferred pharmaceutical preparations were phosphorus in glycerol (known as glycerine in those days), which contained one-hundredth of a grain of phosphorus per millilitre of liquid or tincture. Alternatively, there was tincture of phosphorus, a solution in pure alcohol, which was stronger and contained twice as much.

Elemental phosphorus was still regarded as the proper treatment for certain conditions as recently as the 1920s, and an article printed in 1921, in the *Journal of the American Medical Association*, specifically recommended it to stimulate the formation of new blood cells and promote bone growth. Phosphorus was known to change bone structure, forming dense sections of bone in places. For this reason it had at one time been prescribed treatment for rickets and was still given to enhance the healing of massive fractures. It did not disappear from the *British Pharmacopoeia*, the official listing of all prescribable medicaments, until 1932 and was still available as an over-the-counter remedy in the UK in the 1950s.

The availability of such medicines invariably produced fatalities and, because of the popular belief that phosphorus was a brain tonic or an aphrodisiac, people overdosed themselves. Generally, those who wished to experiment in this way or even to commit suicide used the more readily available forms of phosphorus: match heads or rat poison.

Ashburton Thompson's *Free Phosphorus in Medicine* was a

scholarly work by the leading consultant surgeon to the Great Northern Railway Company of Great Britain and the Royal Maternity Charity, London. This weighty tome not only assumed that phosphorus was a useful medicament but also reported cures that had been achieved with it. Thompson discussed its use in a series of conditions, namely nervous exhaustion (which today we would refer to as a nervous breakdown), melancholia, softening of the brain and hysteria (psychiatric disorders), apoplectic paralysis (stroke), sclerosis of the spinal cord, impotence, migraine, epilepsy, assorted skin diseases, pneumonia, alcoholism, TB, cholera and various conditions of the eye, such as amaurosis (loss of sight due to diseases of the eye, optic nerve or brain damage), cataract and glaucoma. He particularly recommended phosphorus as the best cure for toothache and neuralgia.

According to Thompson, the use of phosphorus had been endorsed by many notable medics down the years and he quoted observations made in the eighteenth century by eminent doctors* and especially the work of Leroy and his treatment for TB. This chorus of approval was to continue in the early years of the nineteenth century with other influential medics† adding their support for its curative powers. With such a weight of authority lending phosphorus its support, it is not surprising that this element was seen as the first-choice remedy for many ailments. Stories were often told of patients being revived by phosphorus when recovery seemed hopeless.

Phosphorus at this time was considered particularly beneficial to the nervous system, although doctors were alerted to its aphrodisiac side-effects and they were aware that in too large

---

* The list comprises Sachs (who published in 1731), Kramer (1733), A. Vater and J. G. Mentz (1751), P. E. Hartmann (1752), F. S. Morgenstein (1753).

† They were F. Bouttatz (who published in 1800), E. C. Jacquemin (1804), J. P. Boudet (1815), J. F. D. Lobstein (1815) and De Lens (1829).

a dose it was poisonous. That, of course, is true, but so were many of the medicines prescribed by doctors, if taken in large doses. The use of phosphorus in medicine declined in the mid-nineteenth century when it was discovered to be the cause of the industrial disease phossy jaw, which slowly ate away the jaw-bone and left suppurating abscesses in the sufferer's mouth (see Chapter 6).

Rather strangely phosphorus returned to favour in the second half of the nineteenth century, especially in France, after a series of articles published by eminent doctors again suggested it was highly beneficial in the treatment of neuralgia and bone con-ditions.

The pharmacopoeias of the time mention many forms in which elemental phosphorus was available. There was phospho-rus dispersed in olive oil, turpentine or cod-liver oil or mixed with beeswax; there was phosphorus dissolved in solvents such as ether, chloroform, alcohol or even carbon disulphide (which smells of rotten cabbage); and there was phosphorated water, soda water or vinegar.

Elemental phosphorus had to be protected from the atmo-sphere, and phosphorated oils were particularly popular. Not all oils were suitable because some of them reacted with phosphorus chemically, especially those we now know to be polyunsaturated ones. Unsaturated oils have a chemical feature known as double bonds, and these make them more reactive than saturated oils. Polyunsaturated oils have two or more double bonds, whereas monounsaturated oils only have one. The latter were favoured as the medium for phosphorus, which is why olive oil was used in Mediterranean countries, where it was known as *oleum phosphoratum*. In the UK, almond oil, which is also mainly monounsaturated, was favoured and the *British Pharmacopoeia* of 1874 gave its formulation as 2 g of phosphorus in 100 g of oil. Cod-liver oil was sometimes used although this is predominantly a polyunsaturated oil.

Thompson's book advocated doses of one-twenty-fourth of a grain of phosphorus when it is given in oil, and this would be a mere 2.5 mg. He noted that even smaller doses were prescribed in other European countries. He preferred to dispense phosphorus as an ether solution, which was to be taken in doses of one-twelfth of a grain (5 mg) every four hours for neuralgia, although he warned that this might produce nausea. A popular form was a much more concentrated solution of one grain of phosphorus in four grains of chloroform, and the instructions for taking this were to add five drops into four grains of ether and pour this mixture into a glass of port or Burgundy wine. When taken as a general tonic, the dose was one-fiftieth grain (*ca* 1 mg). (For children, the preferred solvent was glycerol, with a little peppermint added to make it more palatable.)

Thompson also warned his fellow practitioners to be careful because there was no known antidote for poisoning by phosphorus. He also said that it could cause all manner of side-effects such as nausea, loosening of the teeth, raised temperature, perspiration, itching and bad breath and also that it acted as a diuretic (although the urine was very cloudy). Needless to say there was the embarrassing risk of exciting 'venereal ardour', in which case treatment should certainly be discontinued, unless the patient was a married man who admitted that he had previously been experiencing a lack of activity in this quarter, in which case the treatment could resume. However, Thompson did explain that its use was not universally recognized for this last condition and that some doctors thought that phosphoric acid was better. In order to address this doubt, Thompson dosed himself with phosphorus, first using elemental phosphorus, and when this did not produce the desired effect, he tried zinc phosphide for five days but again failed to get the required response and concluded that it was not particularly effective. He was, of course, right.

Zinc phosphide (Zn_3P_2) was preferred by some doctors

because it was stable in air and tasteless, but its action was the same. This was first made in 1867 by M. Vigier and the standard way of dispensing this was as one part zinc phosphide with ten parts starch powder, and it was sold in packets each containing about 100 mg of the mixture. This would deliver about 10 mg of the phosphide, which was clearly not enough to kill anyone. Thompson was more generous in his use of zinc phosphide but recommended one-third of a grain (20 mg) as an absolute maximum, warning that the first few doses would be likely to cause vomiting, although he added that the treatment should be persisted with, assuring his medical readers that this side-effect generally wore off with subsequent doses.

In the acidic conditions of the stomach the phosphide would be converted to the gas phosphine (PH_3), which might simply be expelled as an odd-smelling burp. At least this form of phosphorus was less likely to damage vital organs, although it might well have played havoc with a few of the body's enzymes.

Dr Lander Brunton in his book *The Action of Medicines*, published in 1897, discussed the effect of phosphorus on the human body. 'The fatty degeneration which is observed after the administration of phosphorus is a toxic action but the same thing may be utilised for its therapeutic action. Fatty degeneration is one of the means by which solid exudations are absorbed in the body and we find that the administration of phosphorus or of arsenic in various forms, tends to help the absorption of certain exudations.' The exudations in question are not spelled out, but it would appear that Victorians were concerned with the human body's ability to ooze various obnoxious substances and felt that these 'exudations' could be channelled through the normal organs of excretion by the action of phosphorus. It was a plausible reason, but in the final analysis there was no scientific support for Brunton's theory, nor ever likely to be, since it was wrong.

Early in the twentieth century, doctors were still prescribing

phosphorus as a general tonic, and in 1931 a Dr G. Coltart wrote
to the *Lancet* about an interesting case. A patient of his had come
to him in 1904 complaining of feeling run down and so the
doctor had prescribed a popular brand of tonic pills that contained both elemental phosphorus and strychnine but told the
patient to stop taking them if the strychnine made him twitch.
The patient had returned twenty-seven years later with an advanced case of phossy jaw, the industrial disease that afflicted
those in the match-making industries, having taken the pills
regularly during the intervening years. Asked why he had taken
the pills for so long, he replied that he continued with them
because they had never caused him to twitch!

Much safer than prescribing phosphorus was to prescribe
the partially oxidized form known as hypophosphorous acid
(H_3PO_2) or its salts, known as hypophosphites. These came into
vogue after 1857 when they were popularized by a Dr Churchill
specifically for the treatment of what was then described as
neurasthenia, a condition characterized by a general lack of
energy. He also advocated their use in the treatment of TB.
Again, there is no reason to believe that hypophosphite was of
any use in treating these conditions, although these compounds
were to persist in patent medicines for more than 120 years,
usually recommended as a general nerve tonic. Earlier in the
twentieth century they were prescribed by doctors for patients
who were suffering from nervous breakdown or clinical
depression or who were convalescing after a serious illness or
operation. The medicine consisted of a mixture of the hypophosphites of calcium, manganese, potassium and iron in a
sugar syrup, with a little quinine, strychnine and chloroform
thrown in for good measure. The only active ingredient was
strychnine, of which there was 1 mg per 8 ml, which was the
standard dose.

The notion of phosphorus as a nerve tonic persists even
today. Sanatogen Tonic Wine has been on the market in the UK

for more than fifty years and it contains 0.62 per cent of sodium glycerophosphate. This is the sodium salt of glycerophosphoric acid, which forms when naturally occurring lectins are hydrolysed and which can also be made by the phosphorylation of glycerol. Calcium glycerophosphate was to be found in medical pharmacopoeia until the 1980s on the basis that this form of phosphorus was more easily assimilated by the tissues, particularly by the brain. There is no evidence to support this belief beyond the long-standing association of phosphorus with mental activity, but calcium glycerophosphate was once prescribed for nervous debilitation and convalescence in doses of 500 mg several times a day.

A communication in the medical magazine *Lancet* (1973, p. 319) reported that four patients with anorexia nervosa who had been hospitalized but had relapsed when they were discharged, were readmitted for a treatment which included the taking of calcium glycerophosphate. When they were discharged a second time it was found that their weight gain was maintained and their eating habits improved. It is hard to believe that their 'cure' was due to glycerophosphate, but that was the implication.

Phosphate treatments

Elemental phosphorus was completely useless, indeed actually harmful, as a form of medical treatment. Other compounds of phosphorus might have been safer but were equally ineffective. For example, George Pearson (1751–1828) converted phosphorus to phosphoric acid and then to sodium phosphate, which he sold as a general-purpose medicament; this acted as a mild laxative, although it could have cured no known disease. This does not mean however that phosphorus-based drugs are without a role in modern medical treatments, and in fact some are particularly effective.

In the 1960s, the detergent giant, Procter & Gamble of Cincinnati, began to investigate whether there were any other uses for the phosphate water-softening chemicals they put into their washing powders. Today, we reap the benefits of these investigations, which have blossomed into a $1 billion industry of phosphate pharmaceuticals. The compounds that were discovered to be medicinally important were the bisphosphonates, chemical analogues of the diphosphates* used in certain wash formulations. Instead of two phosphates being linked through a common oxygen atom, that is $[O_3P-O-PO_3]^{4-}$, they were linked through a carbon, that is $[O_3P-C-PO_3]^{4-}$, the carbon having two other groups attached to it (not shown).

This apparently simple change protects the molecule from attack by water, which can convert diphosphate slowly back into two separate phosphates. Giving bisphosphonates to people in medicines could be fraught with danger since they might well interfere with diphosphate, which is a natural and important part of human metabolism. Indeed, some such bisphosphonates are toxic for this very reason and they block metabolic processes that rely on diphosphate. The most vulnerable molecule is ATP (adenosine triphosphate), which provides the energy to drive most chemical reactions in the body and which can exchange diphosphate for bisphosphonate and thereby be rendered ineffective. However, by carefully choosing the other two groups attached to the carbon, it became possible to remove this kind of toxicity.

As with elemental phosphorus, bisphosphonates are prescribed for bone conditions but now there are good medical grounds for using these drugs and, in any case, they have been thoroughly tested and proven to be non-toxic. Bone is mainly calcium phosphate and, although we consider bone a mineral and therefore stable once it has formed, this is far from being the

---

* Also known as pyrophosphate.

case. Our skeleton reaches its maximum weight when we are around thirty and from then on we lose a little of its mass every year until in old age our bones are porous and liable to break easily, especially at the hip joint.*

Bisphosphonates are used to treat diseases that result in bone wastage of the type that occurs with Paget's disease and certain types of bone cancer.† The most common bone disease is osteoporosis, and bisphosphonates have been approved in many countries for treating this condition. The drugs work by binding strongly to the calcium in bone and this interferes with the action of osteoclasts, the bone cells that break down bone so that its components can be recycled to other sites in the body. The bisphosphonate drugs that are available are marketed under a variety of names such as pamidronate, risedronate, ibandronate and zoldronate. As this last drug can be delivered through a skin patch, it overcomes the major problem of bisphosphonate drugs, which cannot be given orally as they are not readily absorbed through the intestine wall.

Phosphate is an essential component of the human body with many roles in addition to the formation of bone. It is needed for the billions of chemical reactions that take place in our bodies every minute of the day and, most important of all, it forms part of our DNA. We take phosphate in with our food in large amounts, often far in excess of our body's needs, but it is easily filtered out by our kidneys and disposed of in our urine, which

---

* Bone loss is related to gender. It is rarely a major factor in men, except as a result of other illness. In women, however, loss becomes significant at the menopause and is about 1 per cent per year, although this varies according to size, exercise and smoking. Hormone replacement therapy prevents bone loss and this treatment has reversed the trend of increasing numbers of hip fractures in women over sixty-five.

† The drugs have been proved to work using bone density measurements which show that bone loss is halted.

is why this proved to be such a good source of the element in the early days of its discovery.

Patients suffering chronic kidney failure have a problem with phosphate disposal and generally have too much phosphate in their blood, a condition known as hyperphosphataemia. If this is left untreated, it eventually causes osteodystrophy, a painful bone condition in which there are abnormalities in the formation and structure of the skeleton. Even though patients with hyperphosphataemia are put on a low-phosphate diet, it is almost impossible to control the condition in this way and renal dialysis can do little to reduce phosphate levels.

Help for these people should soon be at hand with a new type of drug based on the little-known metal lanthanum. When lanthanum carbonate is added to the food of such patients, it binds strongly with any phosphate released in the stomach and intestines, thereby forming insoluble lanthanum phosphate, which then passes through the gut and out of the body without being absorbed.

Another phosphorus-based drug is Metrifonate, which was originally developed in the 1950s as an insecticide. It was then used by vets to kill parasites such as nematodes in cows and even by doctors to rid humans of the tropical flatworm, *Schistosoma haematobium*, which infests the blood and causes the disease bilharzia. Metrifonate's chemical name is rather a mouthful: dimethyl 2,2,2-trichlor-1-hydroxyethylphosphonate. A new use for the drug is in the treatment of Alzheimer's disease, which it relieves by reducing the activity of acetylcholinesterase (AChE) enzyme in the brain. Although the causes of Alzheimer's disease are still unknown, the condition is linked to low levels of the messenger molecule, acetycholine. This molecule is needed for effective brain function and is removed by AChE when its job is done. Alzheimer's sufferers cannot make enough acetylcholine and their AChE only makes things worse by removing what little there is. One answer is to inactivate the AChE, and this is what Metrifonate can do very effectively, thereby raising the level of acetylcholine in the brain and delaying the progress of the disease.

It is difficult from the vantage point of today's medical treatments, in which new drugs have to pass years of stringent testing before they can be prescribed, to understand how earlier generations of eminent doctors could employ a substance as harmful as phosphorus, and to continue doing so for almost 300 years. They were faced with diseases and epidemics whose causes they rarely understood, and for the treatment of which they had little more than traditional medicaments or blood-letting. The pharmacies of previous generations stocked mercury, arsenic and lead compounds, all highly toxic, as well as plant poisons such as strychnine, so perhaps it was only to be expected that phosphorus would be there as well. In its elemental form it can do us no good, as the safer hypophosphites it is ineffective, and as phosphate it is no different from that which we take in with our food. Only with the emergence of medicinal chemistry was it possible to tailor phosphorus so that it could bring real human health benefits, and the potential now exists for more such healing chemicals to appear in the future.

4. Strike a light

Pack up your troubles in your old kit-bag,
And smile, boys, smile.
While you've a lucifer to light your fag,
Smile, boys, that's the style.
What's the use of worrying?
It never was worth while.
So, pack up your troubles in your old kit-bag,
And smile, smile, smile.

(George Asaf, 'Pack up your Troubles', 1915)

By the time audiences in the music halls were singing this morale-boosting song during the First World War, the lucifer was no more. After eighty years as the world's best-selling phosphorus match, it had been outlawed in 1910, and yet the name lingered on.

In its day, the phosphorus lucifer swept all before it and this kind of match was even praised by the Victorian philosopher Herbert Spencer (1820–1903) as 'the greatest boon and blessing to come to mankind in the nineteenth century'. By the end of that century three *trillion* phosphorus matches were being struck every year.

The campaign to remove white phosphorus from matches had begun in the 1850s when the first safety matches appeared. These relied on the much less dangerous red phosphorus, but they were not as popular as their white counterparts, although in some countries safety matches were the only ones permitted by law. In most countries where match-making was a key industry, the lucifer was king.

It is difficult now to imagine what life was like when most cooking, heating and lighting involved a naked flame. Generating such a flame using flint and tinder could be quite difficult, especially on a cold, damp morning. The phosphorus match did away with the daily struggle to light a fire or a candle and was extremely cheap – 1,200 matches could be bought in London for the price of a postage stamp (one penny) – so it was not so surprising that the benefits of the match were worthy of comment by one of the leading thinkers of the day. The lucifer had ended thousands of years of struggling to light fires, ovens, oil lamps and candles.

Matches had been around since the days of the Roman Empire but were not self-igniting. They consisted of thin strips of wood which had been dipped in molten sulphur at one end. When the tip of a sulphur match was touched against a hot surface, such as a poker or the embers of a fire, it ignited, and a box of sulphur matches could be found hanging on the wall in most kitchens throughout the eighteenth century.

But how did one get a flame on a cold winter's morning when there was no heat to ignite the sulphur match? The answer was to use flint and tinder. By striking a piece of flint against metal, it is possible to knock incandescent sparks from off the flint and, if these were caught on a piece of dry tinder, such as wood shavings or linen threads and gently blown upon they would cause the tinder to smoulder and glow, and from that one could ignite the sulphur match. Provided the tinder was really dry, it was possible to start a fire within a few minutes, but on a cold, damp day it was very difficult indeed and was only achieved at the expense of removing the skin from your knuckles as you struck the flint harder and harder.

Unsurprisingly, many devices were invented in the eighteenth century to create instant flame and some of them involved phosphorus, but the chlorate match was the first way of generating a flame by chemical means.

In 1786, the French chemist Claude Berthollet (1749–1822) experimented with the newly discovered chlorine gas and passed it into a solution of hot potassium hydroxide (KOH), whereupon it precipitated crystals of potassium chlorate ($KClO_3$). He first thought that this stable but powerful oxidizing agent might be a substitute for the potassium nitrate (KNO_3) used in gunpowder, which was in short supply, but the idea came to nothing.

What he did discover though was that sugar and potassium chlorate form an explosive mixture prone to detonate spontaneously while being ground together. When the two powders were mixed as a paste and dried, they became stable but a tiny drop of sulphuric acid would cause the compound to burst into flames, and this chemical reaction was the basis of a new kind of match, the *briquets oxygènes*. These were first produced in France in 1805 and had heads made of potassium chlorate, sugar and gum, and they could be ignited by dipping them into sulphuric acid. Another version of the chlorate match called *Eurpyrion feuerstoffe* was manufactured on a larger scale at a Berlin factory which, by 1825, employed 400 workers.

Another variant was the instantaneous light box invented by Henry Berry of London in 1810. These consisted of a container, one-half of which held chlorate matches, while the other half held a bottle in which there was a piece of asbestos wool soaked in sulphuric acid. A set of fifty matches sold for ten shillings, a price well beyond the reach of most people. Even more expensive were automatic light boxes, which could be put at the side of one's bed to provide light during the night. Pulling a cord attached to the box brought a chlorate match into contact with a drop of sulphuric acid and then moved it on to light a candle.

The danger of chlorate matches was the need to carry around concentrated sulphuric acid to ignite them, but this problem was partly solved by Samuel Jones with his Promethean match, which was named after the Prometheus of Greek legend, who stole fire from the gods. These matches were first produced by Jones in 1829 and sold at his shop, the Light House, 210 The Strand, London. They consisted of a tiny sealed glass tube of concentrated sulphuric acid (about a centimetre in length), around which was wrapped a piece of paper that had been impregnated at one end with sulphur, sugar and potassium chlorate. When the tube was crushed with a pair of pliers, the acid was released and the paper started to burn.

Phosphorus was also an early candidate for instant flame devices, but it came with two problems: problem number one was how to control the flammability, so that this was available only when needed; problem number two was its high cost. The second of these problems was solved in the 1750s when a Swedish mineralogist, Johan Gahn, analysed bone and found it to consist of calcium phosphate. Animal bone was then in plentiful supply and was easily converted to phosphoric acid and thence to phosphorus, a story that is a key part of the development of the phosphorus industry (see Chapter 10).

The first serious use of phosphorus in making an instant flame was the phosphoric tapers which were made in Turin by a chemist called Peyla and were also known as Peyla's candles. They too consisted of a sealed glass tube but were larger, at about 10 cm long (4 inches); along the inside of the tube lay a waxed taper with a frayed end touching a small piece of phosphorus. The user dipped the glass tube into hot water to melt the phosphorus, which was then absorbed by the taper. The tube was snapped open, the taper withdrawn and within a few seconds it would catch fire. In 1791, Peyla's candles were being made in France, where phosphorus manufacture was at a more developed stage. They met a need, but the cost of production negated the cheaper phosphorus which they contained.

Much easier to use was the pocket luminary, which appeared

in Italy in 1786 and was later sold in Paris and London. This was a bottle, the inside of which was coated with phosphorus. When you wanted a light you took a sulphur match, scraped a little phosphorus on to its tip then rubbed this on the bottle's cork, whereupon it would start to burn. The phosphorus bottle had to be kept tightly corked when not in use but, if care was taken, it would last for many weeks. Derepas of Paris was granted a patent in 1809 for 'phosphorus lights' which used a mixture of phosphorus and magnesium oxide, a combination which slowed the oxidation of the phosphorus so the bottles lasted even longer. William Kertland of Dublin came up with a similar idea in 1816, except he mixed phosphorus and camphor.

All these attempts to produce an easy way of generating an instant flame had one drawback or another, and yet there was a fortune waiting to be made for anyone who could produce a cheap, safe and reliable way of doing so. One man who nearly succeeded ran a pharmacy in a small town in the north-east of England.

Friction matches

The first friction matches were made by John Walker (1781–1857) and sold from his shop at 59 High Street, Stockton-on-Tees. Walker, a dapper little man with a cheery disposition and a ready wit, had trained as a surgeon but found it not to his liking and instead became a pharmacist at the age of thirty-eight. In 1825 he was asked by a customer, Alton Norton, to make up a fifty-fifty mixture of potassium chlorate and antimony sulphide as a paste, thickened with a little gum. This formulation was often used as a percussion composition, which means that it would detonate when struck.

A little of the mixture fell on the stone hearth in Walker's workshop and, when he trod on the dried material, it caught fire.

He then experimented by using the paste to tip matches and found they could be ignited by rubbing them on a rough surface. His record books show that at the time he was already making chlorate matches, so it must have been easy for him to make up some of the new type and offer those to his customers instead. Analysis of some of his matches that survived into the 1920s showed the heads to be made of five parts of potassium chlorate, five parts antimony sulphide, three parts gum arabic and one part iron oxide.

Walker's matches were made using thin splints of wood which ensured maximum contact with the head so that, when this ignited, it would quickly set the splint alight. They were best struck by gripping the match head in a folded piece of sandpaper and pulling the match out quickly. If you gripped the match head too tightly, you might only succeed in pulling it from its stick, but with a little practice it became quite easy to light Walker's friction matches, although they always went off with a noisy bang.

On Saturday 7 April 1827, John Walker sold his first pack of 100 matches together with a tin for keeping them in, for twelve pence, and they were bought on credit. (A piece of sandpaper was included in the price.) His shop accounts show that his matches were a modest success and almost 200 sales were recorded in the years 1827–9. Selling only one or two boxes a week was not going to make his fortune, and that might explain why he did not patent them and why he ceased to make them after a few years. Coincidentally, a Yorkshire businessman, Sir Isaac Holden, also produced friction matches much like Walker's in October 1829 and was believed by many to be their inventor, until Walker's shop records came to light seventy years later. Holden's matches were different from those of Walker's though, in that he included sulphur in his formulation for the match head.

In 1828 or 1829 the great scientist Michael Faraday demonstrated a self-igniting match – probably one made by Walker – to an audience at one of his public lectures at the Royal Institution, London, and explained their chemical composition and how they worked. Samuel Jones might have attended the talk because, in 1830, he started making them himself and selling them at his shop in the Strand. He even hit upon a brilliant brand name for them: lucifers. They quickly caught on and soon the name lucifer became the generic term for easy-to-light matches.

The head of a typical lucifer contained roughly equal amounts of antimony sulphide (Sb_2S_3), potassium chlorate ($KClO_3$) and glue.* (The first of these chemicals is the fuel, the second supplies the oxygen.) As yet they contained no phosphorus and so could be tricky to strike, sometimes scattering tiny glowing fragments when they ignited.

Soon Jones' Lucifers were joined by several rival brands such as those of a nearby shopkeeper, G. F. Watts, who named his matches Watts' Chlorate Lucifer matches. Watts solved the problem of transferring the flame from the match head to the match stick by dipping the stick into molten sulphur before applying the match head. When the match was struck, it ignited the sulphur which, in turn, ignited the stick. In 1831, Jones, Watts and several other small match-makers engaged in a vigorous advertising war in the London newspapers which served to boost the demand for matches. In the end, Jones, Watts and Richard Bell of Wandsworth, London, amalgamated to form the

---

* The reaction between Sb_2S_3 and $KClO_3$, which generates enough energy to produce a flame, is given by the chemical equation: $Sb_2S_3 + 3KClO_3 \rightarrow Sb_2O_3 + 3KCl + 3SO_2$. This shows that one antimony sulphide (formula weight 237) reacts with three potassium chlorates (123), so the best combination for a match head is a ratio of 237 of the former to 369 of the latter, which is roughly 2:3, rather than the 1:1 ratio used in the early lucifers.

Bell Match Co. and began to manufacture matches on a large scale. Bell was the main motivator and his innovation was to use match sticks, rather than the thin wooden splints. This made matches sturdier so that they could be struck with some force against any rough surface. Bell also coloured his match heads blue and named his product Blue Bell matches – a brand name that was to last more than 150 years.

Congreves

The introduction of a little phosphorus into lucifer match heads transformed the industry. Who first did this is a matter of debate. Dr Friedrich Moldenhauer of Darmstadt, in a letter dated 1861 to the phosphorus manufacturers Albright & Wilson, says that when he had been a young boy in the early 1800s one of his teachers, a man called Stotzer, had made phosphorus matches. He adds that travelling showmen at fairs used to produce instant flames by rubbing sulphur matches tipped with phosphorus against their stalls. According to Moldenhauer's letter, there was a factory making phosphorus matches in Darmstadt in 1834, but in any event he was certain that phosphorus matches were a German invention.

Others claimed that a J. T. Cooper of London made matches with heads of sulphur and phosphorus early in the nineteenth century but that they were not successful.

None of these early inventions would have amounted to much because the phosphorus they contained would have been quickly oxidized by the air and would thereby have lost its effectiveness. Phosphorus was necessary for easy ignition because of its chemical reactivity – the key to success lay in the way it was used. The answer was not to see it as the *fuel* for the match head, but as the *initiator* of the flame. Only a little phosphorus would then be needed and this could be protected from oxidation by other components in a match head, such as the coating of glue that held it together.

History records three claims to have invented phosphorus matches and they were all made in the early 1830s: by Charles Sauria of France, Jacob Kammerer of Germany and Stephan Rómer of Austria. (The Hungarians also have a claim based on a B. Irinyi, who may, independently, have invented a phosphorus match in 1836.)

The strongest claim appears to be that of a young French chemistry student, Charles Sauria, of St Lothair, who was studying in 1830–1 at the Collège de l'Arc in Dôle, Jura. The head of his match consisted of sulphur, antimony sulphide, potassium chlorate, phosphorus and gum. So why did Sauria not get the credit for his invention? According to one story, he lacked the funds to patent his idea and according to another his discovery was widely boasted about by his chemistry teacher, a M. Nicolle, and the idea was rapidly taken up by others. As a result, Jacob Kammerer of Württemberg, Germany, heard of Sauria's invention and made some matches himself, which he began to manufacture in 1833. The following year, Stephan Rómer, along with J. Siegel, patented a phosphorus match in Vienna.

The French Government, however, eventually recognized Sauria as the inventor of phosphorus matches and, in 1884, when he was in his seventies, the Académie Nationale Agricole awarded him a medal of honour. He died in 1895.

The new matches, which contained phosphorus, were called congreves, after the Englishman who pioneered military rockets (see box). However, in the country of his birth this name did not catch on and the public began to call the new matches by the old name of lucifers – and from now on we will do the same. The main manufacturers of lucifers were in Germany, at Darmstadt, and in Austria, at Vienna, where factories were in production by 1835 and whose products were soon exported all over Europe. (Austria could boast the oldest match factory in the

The man after whom congreve matches were named

William Congreve worked in the laboratory of the Royal Academy at Woolwich, London, where the great arsenal was located. He was commissioned to develop the rocket as a weapon of war and, in 1805, he demonstrated his device, which consisted of a cast-iron tube 3 feet long and 3½ inches in diameter (1 m × 9 cm) with a long guide-stick. The tube could be packed with explosives or incendiary materials. The demonstration so impressed the Prime Minister, William Pitt, that an order was placed for their manufacture. At the time Britain was at war with Napoleon and, in 1807, when there was imminent danger of an invasion from France, 200 of Congreve's rockets were fired from ships at the invasion fleet assembled by Napoleon in Boulogne harbour. A strong wind blew the rockets off course and they fell instead on the town, setting many buildings on fire and doing considerable damage.

This clearly was a much better way of using the new weapon and, the following year, the British launched an attack of 25,000 Congreve rockets against Copenhagen and razed it to the ground. Similar attacks followed on Danzig and Walcheren, and the rockets were even used at the Battle of Leipzig in 1813.

Despite these successes, many in the military were not convinced of their value because Congreve's rockets could be highly unpredictable, sometimes even turning back on themselves and causing damage to the wrong side. After the Battle of Waterloo (1815), they were little used and yet they are remembered every time the US national anthem is played. The red glare of Congreve's rockets is still sung about in the 'Star-Spangled Banner', which records their use in the British attack on Fort McHenry in Maryland during the War of 1812.

world based at Schüttenhofen.) The first US match patent was granted in 1836.

A typical congreve/lucifer match head contained 20 per cent white phosphorus, 15 per cent sulphur, 30 per cent potassium chlorate, 10 per cent chalk and 25 per cent glue.* The formulation of the paste used to make match heads was a closely guarded secret of the various manufacturers. There were many variants: for example, in 1836, Irinyi suggested adding lead dioxide to help ignite the matches, and many others added lead nitrate as well. To begin with, the amount of phosphorus in match heads was quite large, but gradually this declined as its role was recognized to be that of an activator rather than fuel and only a milligram per match head was in fact required. An ounce of phosphorus would be sufficient to make around 25,000 matches. Demand for phosphorus now boosted production to such an extent that there were reports of mass graves on old battlefields being plundered, ostensibly to recover the bones of slain horses.

All kinds of developments now took place. Samuel Jones invented a smouldering match called Fuzees which could be used to light cigars even in a strong wind. They consisted of cardboard soaked in potassium nitrate and were tipped with a phosphorus-containing paste. He also produced Candle Matches, designed to provide a light for about two minutes. These were tiny candles, rather like those used on birthday cakes, again tipped with phosphorus, and when lit they were slotted into a hole in the centre of the match box.

Meanwhile, in the USA other inventors were busy. Alonzo Phillips patented his LocoFoco matches in 1836, which were virtually the same as lucifers, and the right to produce them was bought by Ezekial Byam, the first US match manufacturer. In 1843, William Ashard patented a non-sulphur match by tip-

---

* The average amount of phosphorus in a match head fell to around 5 per cent by the end of the century.

dipping the match stick in beeswax, but this made manufacture expensive. The problem was solved by Charles Smith in 1862, who patented the Parlor match, which was tip-dipped in paraffin wax. His first attempts were unsuccessful because the paraffin wax would not adhere to the wood and both the wax and the match head tended to break away from the match stick when they were struck. He got over that difficulty by heating the match sticks before they were dipped so that the wax could penetrate the wood.

Another problem that needed to be overcome was that of the 'afterglow'. When a match had been used and blown out, the splint of wood continued to glow and, if it was carelessly discarded, could cause a fire. Today, afterglow is prevented by treating the wood with ammonium phosphate, but in the US the problem was solved in 1870 by dipping the match stick in alum or sodium silicate solution. The idea of treating match sticks in this way was not taken up, until it was forced upon match-makers by law.

Afterglow might have been a problem, but phosphorus matches had other more major hazards too: they were easy to ignite and so especially dangerous in the hands of children, and they were poisonous if sucked, which again often put very young children at risk. Not only children were at risk. An item in *The Times* of London of 8 June 1867 told one such tragic story:

The Archduchess Matilda has ceased from suffering. The intended mother of the future kings of Italy, a lady destined to wear a diadem which has not rested on a female brow for centuries ... a Princess in her nineteenth year ... endowed with rare gifts of person, mind and heart, died on Thursday last at 8 o'clock in the morning – of a lucifer match. She inadvertently trod on one which was lying at her feet, as she leant out at the window talking to one of her relatives. Her summer dress was in a blaze before she was aware of it and before anyone could run to her rescue she sank to the ground in an agony of pain, from which only death released her.

Red

Tragedies like the death of Matilda occurred daily throughout the Victorian era, although rarely with such political consequences. Boxes of matches were reported to ignite if they were shaken and houses were set on fire by matches left on window sills which were exposed to the rays of the sun. Much of the misery could have been avoided. Within a few years of lucifers achieving popularity, a safer kind of match was invented and that was made possible by a safer kind of phosphorus: red phosphorus. There is some doubt as to who first made red phosphorus and there are reports from earlier times of phosphorus samples of this colour (see Chapter 2).

White phosphorus exists in units of four phosphorus atoms in a cluster (P_4), but in red phosphorus these clusters link up with one another to give a complex network of atoms. This change confers stability and safety on the element and it becomes neither spontaneously flammable nor poisonous. (Other forms of phosphorus are known – see Appendix.)

One claimant to the discovery of red phosphorus was Emile Kopp in 1844, but his work was overshadowed by that of Professor Anton von Schrötter, Director of the Austrian Mint, who in 1845 produced authentic samples of the new phosphorus. His claim was taken more seriously and he was awarded a gold medal at the Paris Exhibition of 1855 and given the Montyon Prize by the French Academy in 1856 for the discovery. He had produced red phosphorus simply by heating ordinary phosphorus in a strong, sealed glass bulb at 250°C for several days. At the time he considered it of little commercial value – although he did take out a patent on it.

Almost immediately, its potential to create a safer kind of match was realized and the first person to suggest this was Professor G. E. Pasch of Stockholm. Because red phosphorus was stable in air, it could be taken out of the match head and, together with ground glass to act as an abrasive, placed in a panel on the side of the box. It was now no longer possible to ignite the match head by friction alone, it still had to come into contact with phosphorus and that was only to be found on the box. Had Archduchess Matilda trodden on a safety match, she would have lived.

The firm J. Bagge, of Stockholm, was the first to produce safety matches in the late 1840s, but the venture failed because the quality of red phosphorus was unreliable. That was to be remedied a few years later when the firm of Albright & Wilson began to manufacture it in the UK.* Their interest in the new material came as a result of Arthur Albright attending the 1849 meeting of the British Association for the Advancement of Science, which was held in Birmingham, where Professor Schrötter gave a talk on red phosphorus.

Afterwards, Albright introduced himself and negotiated to buy the British rights to Schrötter's patent. He soon discovered that manufacturing red phosphorus was a rather risky business. Often the sealed vessel in which it was being heated exploded due to the formation of phosphane gas from water that was present. To prevent this happening, only the purest phosphorus could be used and this required redistillation until all water was excluded. For safety reasons, it was then heated in a glass vessel surrounded by sand and held inside a metal container. Despite frequent explosions at the works, Albright persisted with the project, but it took three years before he had developed a method

---

* The story of this company is told in Chapter 10.

which worked and which he patented. The method used nowadays is basically the one he devised, which is to heat white phosphorus at 260°C for five days, after which the temperature is raised to 400°C to remove the unconverted white phosphorus, which distils off.

Eventually, Albright was able to offer a reliable grade of red phosphorus for sale, but it was more expensive than white phosphorus. Despite this, it was still profitable to make safety matches because very little red phosphorus was needed and 10 pounds (4.5 kg) was enough for the strike panel on the side of a quarter of a million match boxes.

The first commercially successful safety matches were produced by the Lundström brothers in Sweden, but, when they ordered a *ton** of red phosphorus from Albright & Wilson, who were staunch Quakers, their request was politely refused on the grounds that such a quantity could only be used for making weapons of war. A second letter from Sweden explained why it was needed and the amount was duly supplied. Lundström safety matches were a success and others negotiated to produce them under licence.

The Lundström brothers sold the French rights to M. M. Coignet Père et Fils, of Lyons, for several thousand francs, but later the French firm began legal proceedings against them, claiming that a patent had been taken out by one Böttger in Austria which predated theirs. Links with the fledgling match industry in Great Britain were more harmonious. There, a new company had been formed by the partnership of two other Quakers, William Bryant and Francis May. William Bryant was born to a Quaker family in Plymouth in 1804 and later held a job as a customs and excise officer for fourteen years before setting up as a soap-manufacturer and sugar-refiner. Francis May

---

* An imperial ton is very nearly the same as a metric tonne; 2,240 lbs vs 2,205 lbs or 1,016 kg vs 1,000 kg.

was also a Quaker, born in 1803 in London, where he later worked as a grocer. They were probably attracted to match-manufacture by the rapidly expanding demand – the people of Great Britain were already striking 250 million matches a day and the fact that 80 per cent of these matches were imported at a cost of £60,000 a year presented a profitable opportunity.

The firm Bryant & May was set up in 1852 with £8,000 capital and, true to their Quaker beliefs, it was with the explicit aim of manufacturing only safety matches. The partners entered into negotiation with Lundström and, in August 1855, they were granted a UK patent (no. 55/1854) for safety matches. They bought an old candle factory in Fairfield Road, Bow, on the eastern fringes of London, in which to make them. In 1858, they journeyed to Sweden to the Jönköping match factory at Wester Brunn, where they were wined and dined in great style and donated 1,000 Swedish Riksdals to found a Sunday school for the children of the work people.*

Bryant & May were puzzled by the curious reluctance of the British public to buy their safety matches, whose benefits were clear for all to see. Advertisements were placed in newspapers and these cited the unhappy end of Matilda as proof of the disaster that might be avoided if only people would buy Bryant & May Patent Safety Matches. The campaign was only modestly successful. The trouble was people preferred strike-anywhere matches, whatever calamity they might threaten, and the same was true in most countries. However, Finland, Denmark and Switzerland, which imported most of their matches, passed laws banning lucifers in the early 1870s.

---

* When they enquired a few years later how this was progressing, they were told that only 62 of the possible 112 children who were eligible to attend actually did so, but parents had been warned that if their children failed to register they would lose their jobs and attendance was soon expected to benefit.

In Great Britain, the leading industrial nation of the time, people wanted the familiar lucifers and so, by 1880, Bryant & May were making them as well and were by then among the world's largest producers of such matches. The imperatives of the market place had triumphed over Quaker idealism. In the twentieth century, when they were making sixty billion matches a year, the ratio was still 6:1 in favour of strike-anywhere matches. By then, match-making was automatic, and machines operated by a team of only nine people could turn out fourteen million matches a day.

5. Strike!

1888 was a remarkable year. The British Empire, which already encompassed a third of the world's population, continued to grow with the acquisition of Sarawak, Brunei and North Borneo. Victoria was Queen Empress and the Marquess of Salisbury was her Conservative Prime Minister. Leo XIII was Pope and Grover Cleveland, a Democrat, was President of the US, the twenty-fourth holder of that office. In the Netherlands, Vincent Van Gogh painted his masterpiece *Sunflowers*, while in Russia, Tchaikovsky composed his Fifth Symphony. In the UK, John Dunlop invented the pneumatic bicycle tyre and in the US George Eastman produced the Kodak box camera. The Suez Canal was declared open to ships of all nations by the Convention of Constantinople (now Istanbul), and Jack the Ripper stalked the streets of London at night, murdering and mutilating. In July 1888, an equally threatening event occurred in that city: the girls who worked at the Bryant & May match factory came out on strike – and won.

That it should have happened at Bryant & May's was rather unexpected. As we saw in the previous chapter, Bryant & May had been motivated by the highest of ideals when they went into the match-making business, and despite early setbacks the firm eventually flourished, so much so that they built a new model factory at Bow in the mock-Venetian style of the time. Its façade incorporated their trademark, a Noah's Ark with the word 'security' underneath. There they produced lucifers under the brand names of 'Pearl', 'Tiger' and 'Ruby', as well as their

Patent Safety Matches. What caused the strike was really the prolonged economic depression of the late nineteenth century, which had driven down the price of lucifers to an unbelievable one penny (1d) for a *dozen* boxes, each box containing a hundred matches.

In 1827 Walker sold the first ever box of one hundred friction matches for a shilling, or 12d. At the Great Exhibition of 1851, the price of lucifers was down to 2d per hundred, and most were imported. Within a few years, they were down to 1d, then a ½d. Soon they were too cheap to be sold as individual boxes and retailed in packs of a dozen boxes first for 4d, then 3d, then 2d and, finally by the 1880s, for only 1d. Yet still they could be manufactured at a profit. In fact, the cost of production was a ½d per dozen boxes, which seems inconceivable in what was a labour-intensive industry. Yet to survive when trade was bad against intense foreign competition this had to be so. The UK industry had to compete with imports from the biggest producers of matches, Sweden and Austria, which could supply them at the same price, while in Moravia matches were sold in packs of fifty boxes for 4d, with each box containing 100 matches (= 1,250 matches per penny). Bryant & May had to compete and this was possible only by paying low wages, and that meant employing teenage children or young women. But the worm was about to turn.

Match-makers had always employed large numbers of children. The Children's Employment Commission of 1863 investigated the industry and discovered that more than half the workforce consisted of children aged twelve and under, with some as young as six. Their report included such cases as that of William Needam, aged eight, and Edward Brown, aged eleven, who worked from seven in the morning until seven in the evening with an hour off for lunch. As a result of such revelations the Factory Acts Extension Act of 1864 was passed, bringing the dangerous trades investigated by the Commission within the scope of the existing Factory Acts, which laid down a minimum age of employment and hours of work. As a result the match-makers took on more young women, the alternative source of cheap labour.

Bryant & May had become a public company in 1884 under the chairmanship of Wilberforce Bryant, William Bryant's son.* It employed 3,000 people, of whom the majority worked at home making match boxes, with around 1,300 working in the Bow factory itself. While the original partners were in control of the Bow factory, the workforce had been extremely loyal. So much so that on one occasion they even demonstrated on their employer's behalf against the Government. The demonstration took place in 1871 when a tax of ½d was levied on a dozen boxes of matches. The Government had recently abolished the sale of commissions to army officers and was seeking ways to replace the lost revenue.

Bryant & May and the other match-manufacturers were outraged and began a public campaign to stop the tax, presenting it as a transfer of taxation from the rich to the poor. The match-makers organized a mass rally on 23 April in Victoria Park near the Bow factory and a protest march the following day when their employees walked through the City of London to the House of Commons at Westminster, to present a petition. *The Times* newspaper witnessed the procession, which turned into a near riot:

> A demonstration of matchmakers to petition Parliament against the imposition of the match tax was yesterday attended with riots in the East End of London and a riotous assemblage around the Houses of Parliament. Those who made the demonstration were principally working girls and boys, a year or two in their teens and beyond doubt of the

---

* Bryant had died in 1874 and May was to die in 1885.

working class. They numbered several thousands and were accompanied by men and women of their own class, without any admixture of the usual agitators.

In fact, the powers-that-be made several attempts to break up the march with lines of policemen and used mounted patrols to block the way, but, each time they were confronted, the marchers dispersed into side streets and reassembled beyond police lines. When they reached the Thames Embankment, however, the police were waiting for them in far greater numbers and attacked the leading marchers, throwing their banners and placards into the river. The crowd retaliated with stone-throwing and the march descended into an unruly mob until it was eventually broken up and the protesters driven back towards the East End. Sarcastic letters to *The Times* from spectators who had witnessed the police brutality commended the force for showing such bravery in dealing with a peaceful column of poor working girls. Even Queen Victoria was upset and wrote to the then Prime Minister, Gladstone, saying she was not amused by the tax on matches – and it was duly lifted.*

The price of a match

The process of match-making began with the splitting of poplar wood into sticks twice the length of the finished matches. These were stacked in frames of around 4,000 sticks and clamped tight. Each end of the match stick was tip-dipped in sulphur and then into a molten paste of chemicals known as the composition, or 'compo'. The frames were loaded into ovens and the heads dried

---

* Matches were not taxed for another forty-five years; the tax was levied, in the First World War, at the rate of 2½d per dozen boxes, increasing to 8½d in 1940 during the Second World War and finally to 13¾d in 1949.

for several hours. When dry, the clamps were released and the tipped sticks fed through a machine which cut each stick into two matches. The finished matches were then packed into boxes of 100 and wrapped into bundles of a dozen.

The composition for the match heads was made by melting glue into a small amount of water, then adding to it pieces of white phosphorus to form a thick paste. The other ingredients were then added: the colouring agent (usually potassium permanganate), potassium chlorate, antimony sulphide (or sulphur) and powdered glass. When the whole had been thoroughly mixed, it was poured on to a hot iron slab heated by steam to keep it at the right consistency and maintained at the right depth for dipping the match sticks. A skilled dipper could dip 1,400 frames in a ten-hour working day, making more than ten million matches. However, it was those engaged in the boxing and wrapping of the final product who made up the vast majority of the employees.

The East End of London in the 1880s was poor and overcrowded and most of its inhabitants were out of work. Its two major industries had collapsed. Shipbuilding, which employed 27,000 men in 1865, had sunk in the financial crash of 1866–8 never to resurface, and the silk industry, which had employed 50,000 people earlier in the century, was down to 3,300 employees by 1880. Moreover, there was also a massive influx of Jewish refugees in the area who were escaping from the pogroms of Eastern Europe to the safety of the British Empire. The population of the borough of Bow, where the Bryant & May match factory was situated, more than tripled from 11,500 in 1861 to 37,000 in 1881. Whole families were living in single rooms as houses were let and sub-let, and inevitably this created a pool of labour that kept wages low.

Home working was encouraged by manufacturers not only because it reduced overheads but because those working at home were not covered by the Factory Acts, which specified maximum hours of employment and minimum safety conditions in the workplace. Match-box making was the home-based part of the industry and those employed were paid only 2¼d per gross (144) of completed boxes, out of which sum the home-makers even had to pay for the glue they used. While working at home could be advantageous for the mothers of young children in families where the husband was also in work, for those who had to work at home to eke out a living for the family by making match boxes, life was grim. They were the victims of 'sweating'.*

The appalling nature of these working conditions can be gleaned from the item 'A Christmas Tragedy' which appeared in the *London* newspaper of 30 December 1897. A family of match-box makers had all perished when a fire had broken out and spread so rapidly among the cardboard, paper and glue stored in the room where they lived that they had been unable to escape. Thirty-nine-year-old Sarah Jarvis and nine of her children, aged between eight months and sixteen years, died in the flames. (Mr Jarvis, who worked as a match-dipper, had been in hospital with TB and died the day after the fire.)

Work in the factory itself was mainly done by women and teenage children. 'We find that we can always get as many hands as we require' was William Bryant's comments to the Children's Employment Commission, although he noted that the quality of the workforce varied considerably: 'it is not skilled labour, though some, from practice, will do three times the amount of others'.

The Children's Employment Commission found that employees in the match factories were 'the poorest of the poor and the

---

* The sweating system involved a middleman who farmed out the work. Such a man made his profit, not from the manufacturers who gave him orders and supplied the bulk of the raw materials, but by paying his home workers as little as possible.

lowest of the low' and 'most had no education and were only fit for mechanical employment'. The Elementary Education Act of 1870 improved things. This required all children to attend school until they were thirteen, thereby removing them from the labour market. Now, when they were ready for full-time employment, they were better educated in that they at least knew how to read, write and perform simple arithmetic. The stage was set for a drama of social reform and, cometh the hour, cometh the woman.

'White Slavery in London'

No one now remembers the names of the women who formed the strike committee at Bryant & May. They were Alice France, Kate Slater, Mary Driscol, Jane Wakeling and Eliza Martin. They had their fifteen minutes of glory, changed the world for ever – and were never heard of again. The woman who organized them to act and gave them the support they needed is still remembered: she was the indomitable Annie Besant.

The odds were heavily stacked against the match girls bettering their lot: they were unskilled workers; unemployment in the East End of London was very high; and they were female. That a group of young women could organize themselves and stage a successful strike was a remarkable achievement and their success spurred others to do the same – the modern trade union movement and women's liberation can trace their origins to the courage shown by the match girls.

In fact, the strike of July 1888 was not the first time the match girls had tried to better their lot by taking industrial action. In 1871 and again in October 1885, the girls at Bryant & May's factory had gone on strike because their wages had been cut and also because they wanted their working conditions to improve. On each occasion the strike had quickly collapsed. A pamphlet written by Tom Mann in 1886 pointed out the justice of their claims, noting that the company had paid its shareholders dividends of more than £33,000 and so could afford to remedy some of the worst conditions forced upon the girls. But wages were falling everywhere as trade worsened, and inevitably prices fell.

Mrs Annie Besant was a fiery social reformer with socialist tendencies and boundless energy. She was encouraged by the playwright George Bernard Shaw to join the left-wing Fabian Society when it was founded in 1884.* Annie was a colourful character who was born Annie Wood in 1847. She had been married to a Lincolnshire vicar, the Rev. Frank Besant, but it had been a stormy marriage. When she secretly wrote a pamphlet questioning the divinity of Jesus Christ and had it published, it was too much for him. In 1873, he obtained a legal separation, although she was allowed custody of her young daughter Mabel. Annie Besant became far more notorious a few years later when she published an article on birth control, then a taboo subject, which sparked a court case in which she was tried for obscenity. She was legally declared an unfit mother, so her daughter was transferred back to the custody of her father.

Annie edited a weekly newsletter, the *Link*, which sold for a halfpenny every Saturday and for which she also wrote articles, often inspired by the opinions of her fellow Fabians. At one of their meetings, on 15 June 1888, she heard a talk given by Miss Clementina Black, who spoke with passion about the way women were being exploited. Miss Black urged the formation of a Consumer League, whose members would pledge to buy only from shops which certified that their goods were not produced by sweated labour. In the discussion that followed, she cited the plight of the Bryant & May girls as an example of the worst kind

* She left it in 1889 when she embraced Theosophy and went to live in India, where she led the campaign for Independence.

of exploitation, and a resolution was passed 'that this meeting, being aware that the shareholders of Bryant & May are receiving a dividend of over 20 percent and at the same time are paying their workers only 2¼ d. per gross for making match boxes, pledges itself not to purchase, or even to use, any matches made by this firm'.

Besant could scent a good story and decided to find out more. She visited the Bryant & May factory, where she interviewed some of the match girls. The result was an exposé entitled 'White Slavery in London' published in the 23 June issue of the *Link*. Little did she realize the effect it would have.

The article made some astounding claims, many of which were clearly designed to upset the Bryant & May management, who immediately said they would sue for damages. Besant even found ammunition in the 16-foot statue of William Ewart Gladstone, the Liberal Prime Minister,* which the company had erected near Bow Church and which was inscribed at its base: 'the gift of Mr Theodore Bryant'. Besant claimed that this had been paid for by the workforce, each of whom had contributed a shilling, and that these contributions had not been entirely voluntary. According to the girls she interviewed, they had 'thrown stones at the statue and cursed it, even to the extent of daubing it with their blood'.

But it was not only Besant's inflammatory language that put over her message, it was also the substance of the tale she had to tell; people were shocked when they read of the company's system of punitive fines which were imposed on the girls for minor misdemeanours: 3d for having dirty feet (most of them worked barefoot), 3d for having an untidy bench, 5d for being late and 3d for talking. Those who put matches into boxes were particularly badly treated: if they put 'burns' (spent matches that

---

* He had held office in 1868–74, 1880–5 and 1886, and was again Prime Minister in 1892–4.

had accidentally caught fire) on their workbench they were fined one shilling (12d) and they had to pay the boys who carried the frames of dried matches from the drying ovens to their benches out of their own pockets. The girls who wrapped the filled boxes into bundles of a dozen had themselves to supply the glue and brushes with which to seal the packets.

What also upset readers of the *Link* was the way Besant compared the generous dividends the company paid to shareholders with the miserable wages they paid to their employees. She noted that Bryant & May had paid a dividend of 23 per cent in 1887 and 25 per cent in 1888 and had even paid one as high as 38 per cent a few years earlier. She had also discovered that among the Bryant & May shareholders were fifty clergymen, so she directed an emotional plea in their direction. She explained that they received their 'monstrous' dividends because Bryant & May ran an East End 'prison house' of labour where match girls slaved for ten hours a day, six days a week, for wages of around 6s per week.

Annie Besant described how the girls started work at 6.30 a.m. in summer and 8 a.m. in winter and worked until 6 p.m. They were allowed half an hour off for breakfast and an hour for lunch but otherwise were on their feet all day. She cited the case of two sisters who worked at the factory: one earned 4s per week, while the other who earned 'good money' was paid 8s. However, their combined weekly income of 12s was not take-home pay because of the many deductions made at the factory. The rent on the one room in which they lived was 2s per week and their diet consisted mainly of bread and butter with the occasional pot of jam.

On 7 July the *Link* contained another report from Annie Besant on the plight of the girls, and in the same issue she gleefully announced that the promised libel action had not been forthcoming. In an article entitled 'How Messrs. Bryant & May Fight', Besant revealed that the three girls she had interviewed for her earlier article had now been sacked and had been paid the princely sums of 2s 8d, 3s 6d, and 1s 8d (after various deductions). However, they had received payment only after signing a statement which stated that they were satisfied with their working conditions.

By now revolt was fermenting at the factory and it required only a little more encouragement for the girls to take action. In the 14 July issue of the *Link* Annie Besant reported on 'The Revolt of the Matchmakers' which had taken place outside the newspaper's offices in Fleet Street at 3 p.m. the previous day. Two hundred match girls had marched from the factory through the City of London in order to show solidarity with Annie. The paper also reported that Bryant & May had sacked another girl for refusing to sign a paper saying she had been fairly treated. A group of fifty match girls marched from the office of the *Link* to the House of Commons, where twelve of them were admitted and allowed to lobby Members of Parliament. Meanwhile, production at the factory had ceased as 700 girls had decided to join the strike, while the remainder, around 600, voted with their feet and went hop-picking in Kent, the traditional working holiday of the London poor.

Public response to the affair was phenomenal. Already £400 had been received by the office of the *Link* to start a strike fund, and each day more poured in from readers. The all-male London Trades Union Council took a rather remarkable step and agreed to support the girls until the strike was settled, saying that 25,000 working men would turn out if necessary to protect the girls should they be threatened in any way.

The company directors, Frederick and Wilberforce Bryant, denied the allegations made by the *Link* and again threatened a libel action but it was a contest Annie Besant declared she would welcome. She wrote letters to other London newspapers about the girls who had been dismissed for talking to the press, and these letters were published by the *Pall Mall Gazette* and the *Star*.

In the letters she appealed for readers to send money so that the girls might be paid 6s a week until they found work.

The march to Fleet Street, where all the London newspapers were located, had gained excellent publicity for the girls. The *Star* newspaper sent a reporter to talk to them and he too gathered information about their working conditions and their demands. However, the *Star* was the only large-sale daily paper to support the strike. The Bryant & May management had most of the press on their side and newspapers published veiled threats against the ringleaders, including the assertion that the strikers were only making things worse for themselves. Frederick Bryant gave an interview to *The Times* which reported his opinion that 'The women in the Victoria Factory were the ringleaders in the strike ... and he should deal with those in that factory in a way that would make an example of them to the others, probably by refusing to take any of them on again.' *The Times* commented that 'it was a pity that the match-girls have ... been egged on to strike by irresponsible advisers. No effort has been spared by those pests of the modern industrial world, the Social Democrats, to bring the quarrel to a head.' The paper believed that the strike was bound to fail and that the outcome would be yet more unemployment and a reduction in the already low wages of female labour in the East End of London.

Nevertheless the *Link*, the *Star* and the *Pall Mall Gazette* asked their readers to contribute to the strike fund and yet more money poured in with two MPs, F. Wootton Isaacson and W. A. McArthur, leading the way with donations of £2 2s each. Another MP, Charles Bradlaugh, a friend of Annie Besant, asked questions in Parliament. At an open-air meeting held in Mile End, at which Annie Besant, the Rev. Steward Headland and Herbert Burrows spoke, a collection raised a further £5 12s. A similar meeting in Regent's Park produced £4.

Every day the *Star* revealed more about the conditions in the two Bryant & May factories, confirming the girls' claims that all kinds of deductions were made from their wages. Even *The Times* discovered a case of one girl who had been fined more than 2s one week, equivalent to two days' work. More importantly, an independent survey conducted by a group from Toynbee Hall, a respected charity based in the East End, verified the charges that Annie Besant had made against the company. What was also somewhat embarrassing for the directors was that the Bryant and May families were active supporters of the Liberal Party, whose members were broadly in sympathy with the strikers, and up and down the country Liberal Clubs were collecting for the strike fund.

It was finally all too much. Bryant & May conceded defeat on 17 July and on the 18th the girls returned to work. The strike settlement terms were negotiated between the directors of Bryant & May and the London Trades Council acting on the girls' behalf and were announced to the girls themselves at the Great Assembly Hall, Mile End Road. They were published in the *Pall Mall Gazette* on 18 July:

1. All fines were abolished.
2. No deductions for paint, brushes and stamps.
3. The boys who carried the frames of dried matches to the girls would be paid under a different scheme, rather than by the girls.
4. The firm would purchase wheel barrows for carrying completed consignments of boxed matches to the warehouse (instead of the girls having to carry these on their heads).
5. The packers would not be deducted 3d. to pay for the packing paper to be brought to them, they would be allowed to fetch this for themselves.
6. The girls were allowed to form a trade union.
7. The company provided a separate room for the girls to eat their meals.

8. No girl was to be penalised for taking part in the strike and sacked girls were given their jobs back.

Bryant & May had agreed to all the strikers' demands because these were really quite modest and, while they served to increase the take-home pay of the girls and made their working conditions less burdensome, they hardly affected company profits. Bryant & May continued to be the largest match-maker in the British Empire and were still able to pay substantial dividends to their shareholders. Annie Besant declared herself highly satisfied with the settlement terms, which, she said, far exceeded her expectations, so she recommended that the girls accept them, which they did. On 21 July the *Link* announced 'Complete Victory!' and Annie Besant was justifiably proud of what had been achieved in so short a time.

Frederick Bryant sought to have the last word in the dispute and, in a letter to *The Times*, he again rebutted the allegations of Annie Besant and bemoaned what had happened: 'our work people were thoroughly in accord with us, until the socialistic influence of outside agitators disturbed their minds'. Meanwhile, the girls founded the Union of Women Matchmakers, electing Annie Besant as their secretary and her colleague, Herbert Burrows, their treasurer. By November that year they had 600 members, and ten years later helped their sister workers at the Bell Match factory when they too came out on strike.

A sort of calm returned to Bow, only to be broken by the start of the Jack the Ripper murders that summer in the East End of London, which terrorized the area for the rest of that eventful year.

The rise and fall of the Match King

While Bryant & May and the rest of the British match-manufacturers struggled gainfully on in the face of the worsening economic climate, their cousins overseas looked for alternative ways
to survive. The largest match-producing country was Sweden,
and match-makers there were well placed because their country
has a plentiful supply of mature aspen trees, which yield the ideal
wood for matches. One tree will make more than a million
matchsticks but, even so, tens of thousands of trees a year were
needed to meet the demand. This benefit was important to the
industry, but equally important was the entrepreneurial skill of
Ivar Kreuger, the man who came to be known as the Match
King.

The founders of the Swedish match industry were the Lundström brothers, Johan and Carl, who began to make matches
during the 1840s in a small rented workshop in Jönköping.* Carl
was the businessman and it was thanks to his financial acumen
that their fledgling company flourished. Indeed, it grew so
quickly that within two years they were able to build a new
factory, which is now better known as the famous Match Museum. In 1864, the firm took on a mechanical engineer, Alexander Lagerman (1836–1904), who designed the first automatic
machine for the industry. The two brothers and Lagerman
worked well together and the company expanded year by year.

Throughout the second half of the nineteenth century the
demand for matches steadily increased and companies started up
all over Europe and, eventually, around the world. As competition became fierce matches became cheaper and cheaper, but
the Swedish match industry slowly began to dominate. At the
start of the twentieth century, there were two large match

---

* See Chapter 4, p. 79.

combines in Sweden: one centred on Jönköping and the other was owned by Ivar Kreuger. He amalgamated both combines into one giant organization in 1917 which became the Swedish Match Company.

Kreuger was born in 1880, the eldest son of the owner of a small match factory in Kalmar, Sweden. After training as a mechanical engineer at the Royal Technical University of Stockholm, he worked for a while in the booming US construction industry before returning to Sweden in 1908, when he teamed up with Paul Toll to form Kreuger & Toll, a company which specialized in the new construction technique of reinforced concrete. The company flourished, but Kreuger found there were better ways of making money and turned his undoubted business skills to modernizing his father's business, taking over other smaller match factories along the way until his combine, the United Match Factories, rivalled that of the other Swedish matchmaker, the Jönköping-Vulcan Trust, and their amalgamation became inevitable. At each stage Kreuger issued more new stock and raised money on the basis of his holdings in the highly respected firm of Kreuger & Toll.

Next he sought to expand his match empire overseas and soon had captured most of the Asian, British and German markets, mainly using loans from willing Swedish banks. He acquired stock in the US Diamond Match and, during the 1920s, his fortune grew until he was credited with being one of the world's wealthiest men. His speciality was in taking over the Government match-making monopolies which existed in countries such as France, Estonia, Yugoslavia, Romania and Ecuador. He would choose his moment, securing the deal in return for massive loans of which the obliging governments were often in dire need. The French Government was so grateful for his help that it awarded him the Grand Cross of the Légion d'honneur.

Kreuger even appeared to weather the Wall Street crash of

1929 and was certainly able to help the German Government with a loan as it struggled with mass unemployment. There is evidence that he was willing to fund Hitler, whom he considered a bulwark against Soviet Russia, one of the few countries that had refused to let him take over its match-making industry. By the early 1930s Kreuger owned 250 match factories in forty-three countries and had cornered 75 per cent of the world's match trade.

At the time it appeared that match-manufacture might be one of the few industries that would escape the Depression of the 1930s because matches were an everyday essential item, even among the poor and unemployed, but in fact Kreuger's empire was tottering. The money he had borrowed could not be repaid and his massive movement of funds and loans between countries was a desperate attempt to hide the truth. To escape his creditors he resorted to printing forged Italian Government bonds in an effort to convince them that he could cover their loans, but the final act came on 11 March 1932 when he fled to his Paris apartment, spent a last night in the arms of his current mistress, a young Finnish girl, and shot himself the following morning.

Surprisingly the Swedish Match Company survived, mainly because Kreuger had been moderately honest with its business affairs. It was Kreuger & Toll which bore the brunt of his financial scams with a £40 million deficit (equivalent to around £4 billion, or $6 billion, in today's money). Yet the Swedish Match Company was to have an equally chequered career in the second half of the twentieth century.

The afterglow of the humble match

At the start of the twentieth century Bryant & May's greatest rival was the Diamond Match Company of America, which had bought itself into the British market by the acquisition of a small match-making firm near Liverpool. There it manufactured its Puck and Captain Webb brands of matches.* They transformed the match factory by installing a new continuous match-making machine that could cut and dip 600,000 matches an hour, and as a result Puck and Captain Webb matches were sold at prices that Bryant & May simply could not match. Bryant & May responded to the threat by buying the Diamond Match Company in 1905. They also cut competition by taking over other British match-makers including Moreland & Sons of Gloucester, whose best-known brand was England's Glory. The company was famous for the excruciating jokes printed on their match boxes:

> *Child:* 'Mummy, Mummy – Daddy's going out!'
> *Mother:* 'Then hurry up dear and put some more petrol on him.'

In 1926, Bryant & May combined with the major match-importers J. John Masters and amalgamated with the Swedish Match Company to form BMC, the British Match Corporation. The BMC survived the Kreuger affair and after the Second World War, as cigarette smoking became universally popular, it even invested in new facilities for some of its factories, although not for the one at Bow. In 1973 BMC merged with Wilkinson Sword, who manufactured razor blades and garden shears, to become Wilkinson Match. Nevertheless, the market for matches began to decline as home-owners turned from coal fires and gas cookers to central heating and electricity, and as people started to give up smoking.

Soon Wilkinson Match was taken over by Allegheny of Pittsburgh, but that company collapsed and was bought out by Stora, a Scandinavian maker of floor coverings, which sold it on

---

* Captain Matthew Webb was the first person to swim the English Channel, which he did in twenty-two hours on 24–25 August 1875, and became a popular hero.

to an Italian bank and a Nigerian chieftain. Finally, in 1992 it was bought by Procordia, a consumer products company owned by Volvo.

The Bow factory of Bryant & May finally closed its doors in 1979. In 1988, exactly a hundred years after the match girls' strike, it was bought by a development company which planned to convert it to housing. Despite that company collapsing in the recession of the early 1990s, the project was eventually completed and the factory emerged as Bow Quarter, consisting of hundreds of upmarket warehouse-style apartments for professional people and City workers, together with various facilities such as security guards, fitness centre, swimming pool and convenience stores.

The Bryant & May subsidiary struggled on and launched their 'Green Check' campaign in March 1992, trying to show how environmentally friendly they had become. They had reformulated the heads of strike-anywhere matches to eliminate sulphur, which was accused at the time of causing acid rain, and had replaced it with ferrophosphorus, a by-product of phosphorus manufacture. They had removed the zinc oxide filler, said to be toxic and harmful to fish, and replaced it with limestone (calcium carbonate). They had cut down the amount of animal-hide glue in their matches and their match boxes were now made from recycled board. They still used potassium chlorate, although they claimed to be considering 'other options'. But their good intentions were to no avail. The British, like other Europeans and the Americans, had simply stopped buying matches, not because they were environmentally unfriendly but because they were no longer needed.

Other parts of the world still rely heavily on matches, but there has been no significant change in their design for nearly 100 years. The match stick is still best made of aspen or poplar wood treated with a solution of ammonium phosphate and phosphoric acid to prevent afterglow. The match sticks are tip-dipped with paraffin wax before the head is applied. For a strike-anywhere match, the head consists of tetraphosphorus trisulphide and potassium chlorate, bulked out with a neutral 'filler' such as zinc oxide or plaster of Paris, plus an abrasive such as powdered glass and a colourant, all held together with glue.

In the developing world, match-making is often carried out in small factories employing only a few people: in South India there are 500 such firms, each serving its local market, and they require relatively little in the way of raw materials, apart from a few simple chemicals. Nevertheless, it is their demand for potassium chlorate which still accounts for most of the 50,000 tons of this chemical that are produced each year.

6. The cost of a box of matches

> Top grade selectable
> Hardly detectable
> Phosphorus, phosphorus.
> Taste is more subtler and
> Spreads just like butter – grand
> Phosphorus, phosphorus.
>
> Our special beauty cream
> We look a proper dream –
> For we are minus a jaw.
> Guv'nors don't charge a fee,
> Give it away for free,
> Phosphorus, phosphorus.

This comes from Act I scene 1 of the musical *The Matchgirls*, by Bill Owen and Tony Russell, and is sung by the employees of Bryant & May. The line in the lyrics which seems rather incongruous is 'For we are minus a jaw', but this is not just poetic licence – a few of the match girls really did suffer this terrible fate.

One of the saddest sights in the Odontological Museum of the Royal College of Surgeons of England is the display case that preserves the jawbones of some of the match girls. It is impossible to imagine the pain they suffered as the corrosive condition, known as phosphorus necrosis, eroded their teeth and gums to such an extent that it sometimes ate its way completely through the bone. One sufferer described the pain as like the continual rasping of a file.

Accounts of their condition make dreadful reading. In the case of one such match girl – the hospital records give only her initials, M.A.D. – she was eighteen when she got a job at Bryant & May's factory in 1887. She may even have been the Mary Driscol who was one of the leaders of the match girls' strike the following year. To begin with she worked as a cutter-down, feeding the machines that cut matches, later she worked as a box-filler. Life was hard but the wages she brought home, especially after the strike, must have been a welcome addition to her family. She married when she was twenty-two and her first baby, a boy, was born a year later.

When she was twenty-four one of her wisdom teeth started to ache and, eventually, it got so bad that she visited a local dentist. He could see it was too decayed to be filled and pulled it out. The cavity it left behind refused to heal and, after a few days, it began to discharge a vile-tasting pus which gradually became worse until sometimes it oozed from her nose. Still she worked on at the match factory. Eventually, her jawbone became exposed and one day a piece of it came away. The pain in her swollen face was indescribable, but she knew what the matter was: she had phosphorus necrosis or phossy jaw, as it was commonly called, and this was confirmed by the company doctor, Dr Garman. He sent her to the London Hospital, whose records tell her story:

M.A.D., aged 30 years, twelve years as a box-filler and cutter-down.

History About six years ago she had a carious wisdom tooth, which was extracted. An abscess formed inside the mouth over the palate bone. This discharged and a small piece of bone came away.

Present Condition She has now a purulent discharge from the left nostril but no other symptoms of necrosis.

Family History Fair. Father died at 59, mother at 46; two

brothers alive at 32 and 27 years. She has had three children. Her first child is alive and healthy, aged seven and of the others born since her own illness, one lived two years and died from infantile diarrhoea and another baby was stillborn, said to have been strangled in its membrane.

What finally happened to M.A.D. is not recorded.

Phossy jaw was nothing new. It had been a hazard of the phosphorus-manufacturing and match industries for more than thirty years. The first victim was a young woman, Marie Jankovitz, who worked in a match factory in Vienna, and the case was publicized in the medical literature of 1838. Chronic phosphorus-poisoning is slow to occur and the average time between first exposure and onset of the disease was about five years. Of those exposed to phosphorus vapour, only about one person in twenty went down with the disease. Nevertheless, the condition was so horrific that, even when only a few cases had been notified, the Austrian Government set up a commission to enquire into conditions in the seven match factories that then existed in the capital. The commissioners suggested ways of improving the factories, but they were never implemented and cases of the disease continued to occur at the rate of around ten a year.

The first case of phossy jaw to come to the attention of the medical profession in Britain was in December 1846, and in 1847 a few people with the disease were treated at Guy's Hospital, London. The eminent doctor Sir Samuel Wilks noted the excessive suppuration of sores and the exfoliation of the bone, a term used to describe the way in which it broke away. From that year on, there was to be a steady trickle of patients with the disease to hospitals in London and, in the Bryant & May factory, there were fifty-one cases in twenty years, of whom nine died, and the average age of death was thirty-two years.

French doctors were another group to take phossy jaw

seriously and, in 1858, M. A. Chevallier and A. Poirer reported their findings in the *Journal de Chimie Médicale*, describing in graphic detail the slow progression of the disease, from the loss of teeth to the gaping holes which oozed the most offensive-smelling slime. They had observed sixty cases and recorded that half of the people with the disease committed suicide, unable to tolerate their miserable existence any longer. Even those who bore it with fortitude still stood a one-in-five chance of dying from the condition, which undermined their general health.

An article in the British trade paper *Chemistry News* in 1861 warned that match-manufacture was hazardous and that phossy jaw afflicted those whose job it was to dip, dry, cut and box the matches. (One reader responded with the suggestion that phossy jaw might be prevented by the smoking of *arsenical* cigarettes, although it is not clear why these would seem to be an appropriate treatment, nor for whom these rather curious objects were normally intended.) The incidence of phossy jaw among workers in Britain was around 1 per cent, which was about average for this industrial disease, whose incidence varied from 0.3 per cent in the US to 3 per cent in the Thuringian district of Germany.

A typical case of phossy jaw was described in the first issue of *St Bartholomew's Hospital Report* in 1865 by Dr T. Smith. This was the leading London hospital of the day and it was to Bart's that a patient was referred whose face was greatly swollen and who was no longer able to eat solid food, the result being that he was seriously undernourished. He was a thirty-five-year-old lucifer match-maker. The report read:

Extending from ear to ear along the line of the jaw was a chain of ulcerated openings, from which there was a profuse discharge and through any of which a probe reached dead bone. Inside the mouth ... was seen bared soft parts in its whole extent, the bone being rough and brownish-black. The gum gaped widely away from the dead jaw and had receded so as to leave it above the natural level of that bone; a probe could be passed easily either side in front or behind the bone towards the sinuses in the neck.

The man was operated on under chloroform and his jawbone removed by cutting through it in the middle and then dragging the two halves out: 'considerable force was required to detach the bone from its connections . . . and the dead bone came away completely denuded of soft parts . . .'

Surprisingly, the patient made a good recovery and, as his wounds healed, the swelling in his face went down and six weeks later, when he was discharged from the hospital, there were already signs that a new jawbone was starting to form. However, the story does not have a happy ending. He went home and celebrated his recovery to such an extent that he died from inhaling his own vomit while in a drunken stupor. When his body went for autopsy, the newly formed jawbone was removed for examination and preserved for posterity in the Museum of Morbid Anatomy at St Bart's.

Conditions in Victorian factories were notoriously bad. When a Dr Mitchell visited a match factory in 1843, he wrote that the atmosphere within was exceedingly disagreeable but was not surprised that no one complained because the children could earn twice as much working there as they could at other occupations, some taking home seven shillings a week. In 1862 the Queen's Privy Council in London had commissioned a report into phossy jaw, but nothing came of it. The report recommended better ventilation in match factories, facilities for washing hands and separate rooms where meals could be eaten, but these were not made enforceable by law. Little was done to prevent this terrible disease and, as the teeth of most workers were in a deplorable condition, those in the match factories were particularly vulnerable. Dental hygiene had taken a rather heavy blow in the middle of the nineteenth century, due in part to cheap, sugar-based confectionery and the popularity of sweetened condensed milk. The latter had first been advertised by Gail Borden of Burrville, Connecticut, in May 1858 and was marketed in Europe by the Anglo-Swiss Milk Company in 1867. In Britain it was soon to be found in every working-class home.

Phossy jaw was also known by those in the industry as 'the disease', 'the flute' or 'the compo', this last name being derived from the 'composition' paste that was used to make match heads. The component responsible for the disease was, of course, phosphorus, of which 60 tons per year were used to make matches.

Andrea Rabagliati, Honorary Surgeon to the Bradford Infirmary, in Yorkshire, was an expert on phossy jaw and in his book *Health and Disease in Industrial Occupation*, written in 1880, he described the disease in graphic terms:

> For a certain period of time, differing in different cases, the operative feels pretty well. After this a certain disposition to a sort of influenza, accompanied by sneezing, is set up; and the operative suffers at first occasionally, but afterwards more frequently from tooth-ache. These pains in the teeth, which may attack sound as well as diseased teeth, are of grave import to workers in phosphorus, particularly when they do not yield to ordinary treatment. After a long exposure to the agent, the pain is no longer confined to the teeth but spreads to the whole of the upper and lower jaws and frequently strikes even over the whole face and the neighbouring portions of the throat; the throat glands swell and become painful, the gum becomes inflamed and the cheek participates in the swelling and grows hard and tense. The gums are for the most part, soft and elastic, particularly in the upper jaw, while on the lower abscesses frequently form which discharge loathsome and putrid matter. The teeth on the affected side, whether these were decayed or not, become loose and either fall out of themselves, or may easily be

pulled out by the fingers. More abscesses now form on the gums, which lose their bright red colour and become livid and are so undermined by the matter as to suggest sometimes the appearance of a sieve, through the holes of which an offensive discharge wells up. In some cases the whole gum, the bone, the tissue composing the cheek and even the throat, have been attacked and eaten away by this terrible malady.

Rabagliati estimated that around 10 per cent or more of those employed in the industry would eventually be affected by the disease in some form or another, although this depended on the ventilation within the factory. Those who suffered the worst cases of the disease had to have operations to remove their decayed jawbones and have a false jaw and teeth inserted in its place. What made the disease so distressing was the fetid discharge in the victim's mouth, which was so objectionable that even carers and nurses would try to avoid those with phossy jaw if possible.

The records of the London Hospital, to which many of the Bryant & May workers with phossy jaw were sent, show the awfulness of what happened. Patient no. 597 in the year 1879 was a thirty-year-old man, initials A.R., who was treated by Dr Hughlings Jackson. For twelve years A.R. had worked as a matchdipper but three months previously he had gone to the dentist complaining of pain in the gums and teeth of his upper-left jaw and three teeth had been extracted. This produced no relief and he went to see a doctor, who sent him to the hospital.

There, he was examined more thoroughly and it was noted that his left cheek was swollen and his gums were very tender. Many of his teeth were loose and the surrounding gums were oozing pus. In August and September he was operated on three times by surgeons, who removed his teeth and the affected bone of his upper jaw. In September A.R. suffered pain and swelling on the right side of his face and mouth, and more bone was removed. For the next five years he was ill, with his upper jaw discharging pus continuously, until in 1884 he could stand it no more and was readmitted to the hospital, where in June, July and August of that year he had four further operations to remove parts of his lower jaw. But improvement was only temporary and, in November that year, it was removed entirely. His hospital record card contains no further information which suggests he was eventually sent home, although it is unlikely he survived for long. His lower jaw was preserved for the London Hospital Medical College museum.

The 1898 Government Report

The Queen's Privy Council may have been unable to do anything about phossy jaw, but the problem did not go away, although it was another thirty years before the Government took any action and this was only in the way of setting up a panel to look into the matter. It was convened in 1898 and issued its findings the following year: *The Report on the Use of Phosphorus in the Manufacture of Lucifer Matches.* Written by T. E. Thorpe, T. Oliver and G. Cunningham and published by Her Majesty's Stationery Office, it begins, rather ingenuously, with the words:

> In the early part of 1898 the attention of the Home Office was specially directed to the danger attending the manufacture of lucifer matches ... It was discovered that certain cases of phosphorus necrosis among work people had been intentionally concealed and that others had escaped record.

Clearly it had not been the Government's fault that the perils of phossy jaw had been overlooked. Having absolved the powers-that-be from the charge of not taking action earlier, this remarkable document then revealed the scale of the problem. It is packed with statistics about the match industry at the time.

In its pages we find that twenty-five match-makers were employing 4,152 people, of whom 1,070 were male and 3,082 female. Of this workforce, 1,492 were aged under eighteen but only two were under fourteen. Those who were employed in factories making matches numbered 3,134 and, of these, 1,521 were directly exposed to the fumes of phosphorus, some more so than others depending on whether they were engaged in dipping, drying or boxing. Most match-making went on in the East End of London, where 2,256 workers were employed.*

The panel members had spared no effort in getting at the truth and had visited eight European countries as well as obtaining information from the US, where the Diamond Match Co. employed 2,000 people and produced 85 per cent of all matches used in North America. They noted that this company had set up a match factory in Liverpool employing 337 workers where the main processes were automated, using machinery imported from the US. Moreover, the factory was lit with electric light and enormous fans ensured that the five-storey building was well ventilated by changing the air every four minutes; it was even protected by self-closing iron fire doors and an automatic sprinkler system. The twenty-four automatic match-making machines each turned out six million matches a day.

As a result they found little risk to workers there – so much so that this factory was exonerated in the report and its employees excluded from the statistics of those at risk. (The company even offered to pay for false teeth if their employees went down with phossy jaw.) In other factories in the UK, things were far less satisfactory. Employers had disregarded the existing rules

---

* These were at factories in Bow (Bryant & May), Bromley-by-Bow (Bell & Co.), Old Ford Road (Palmer & Son), Stratford East (B. Daniels) and Lamprell Street (Salvation Army). The other towns with large factories were Gloucester (Moreland & Sons), Swansea (Swansea Wax Vesta Company), Liverpool (The Diamond Match Co.) and Glasgow (Mitchell & Co.).

regarding working conditions, had not provided proper dental facilities for detecting phossy jaw in its early stages and had simply ignored – and even concealed – cases of the disease.

The statistics Cunningham gathered for the years 1894 to 1898 show that the disease was rare but getting no rarer. There were two cases notified in 1894, four in 1895, seven in 1896, seven in 1897 and ten in 1898. Some contracted the disease long after they had left the industry, but they could not be taken into consideration. Of the total of thirty cases, sixteen were men, most of whom worked as dippers and three of whom died of the disease, while fourteen were women, most of whom worked in the boxing department and none of whom died.

Those most at risk of contracting phossy jaw were the men who made up the compo, which contained around 5 per cent phosphorus, and those who dipped the matches into it to form the match heads. Chemical analysis of the vapour arising from the molten match paste showed it to be a mixture of 70 per cent phosphorus pentoxide (P_2O_5), 10 per cent phosphorus trioxide (P_2O_3) and 20 per cent elemental phosphorus. The fumes in match factories were found to be contaminated with 0.02 to 0.12 milligrams of phosphorus per 100 litres of air. Today we would record this in terms of 0.2 to 1.2 milligrams per cubic metre. A match dipper would have been exposed to the upper limit for as long as ten hours a day.

The panel were puzzled as to why breathing phosphorus fumes should have such an adverse effect on health. Phosphorus was a tried and tested medicine, much prescribed by doctors for all kinds of ailments, so why should it produce this terrible disease? In the words of the report:

In whatever way the phosphorus may be supposed to bring about . . . changes in the jaw, there are difficulties in the way of accepting the view that the poison is merely conveyed through the carious tooth to the bony tissue . . . It seems

reasonable to believe that the long continued absorption of
a powerful drug, like phosphorus or certain of its oxides,
must produce some effect upon the system . . .

The panel had been told on their trips to Europe that the
bones of those exposed to phosphorus fumes appeared to be
more fragile and that when these were broken they also took a
long time to heal. The medical opinion in those countries was
that a condition existed which they called *phosphorism* and
which was a weakness in some people, making them particularly
susceptible and also that this was especially true of those suffering from TB, syphilis or anaemia, or were under-nourished
or alcoholics. The panel believed that phossy jaw could be
prevented by improving working conditions and paying attention to personal hygiene among staff, such as promoting regular
gargling with antiseptic mouthwash. The report listed fourteen
measures that might put an end to the problem and they suggested the Government might follow the line taken in Russia,
where the Government had taxed phosphorus matches but not
safety matches, to encourage the switch from one to the other.

The panel also looked at what was happening in other
countries and found a wide variety of laws and working conditions. In Holland there were 570 people employed in match-
making but the industry was fully mechanized, including the
dipping process and box-filling. In Denmark match factories
employed 447 workers but they made only safety matches,
because lucifers had been banned by law since 1874. In Norway
the trade engaged 600 people and there too there were cases of
phossy jaw. While Sweden had the largest match-making industry, employing 5,500 people in twenty-seven factories, many of
these were highly automated and phossy jaw was rare. The
industry was geared up to exporting three-quarters of all matches
produced, of which the majority were lucifers. The Swedish
Parliament had debated the 'phosphorus problem' on a few

occasions, but the reason for wanting a ban on its use was not phossy jaw but the large number of suicides who killed themselves with phosphorus.

Germany had a flourishing match industry, although it was generally in small factories, of which there were ninety, and there were about eight cases of phossy jaw per year on average. In neighbouring Austria–Hungary there was a similar number.

By 1899, Bryant & May were leaders in trying to relieve the effects of the disease. Workers with phossy jaw were retired on a pension of twenty shillings a week for women, twenty-nine shillings for men. A fund for the families of those who died of the disease was paid for by voluntary subscription of workers in the factory, which raised £100, and the company added £100. They employed Dr Garman, a local physician and resident of Bow, to monitor the workforce for cases of phossy jaw and treat those affected. Workers at the factory were given free dental treatment and those who had teeth extracted were granted sick-leave until their gums healed. One woman who had worked at the factory for five years went down with the disease and was paid £1 a week for the thirty-one weeks that she was off sick until she was cured.

There was another factory in Bow where matches were made: Bell & Co., which had been founded in 1832. What puzzled the Medical Officer of Bow was that there had been only one case of phossy jaw at Bell's in all the years of its existence and that was only a mild case, whereas at Bryant & May's there had been many, in spite of that company's endeavours to prevent it. The reason was that at the Bell & Co. match factory the mixing of the compo was done on the roof of the building, as was the dipping of match sticks, so the noxious fumes were easily vented away. Bryant & May's original single-storey factory had been converted to a multi-storey building in 1874, but the process of mixing the compo and dipping remained where it had always been, on the ground floor, and, despite attempts to install 'scientific ventilation', the fumes of phosphorus permeated the building.

In 1892, match-making that used white phosphorus had been classed as a dangerous trade and employers were required by law to introduce special ventilation around the dipping tables. This consisted of a pipe where the operator stood which was perforated with a row of holes; from these fresh air flowed across the table to an extraction hood at the opposite side which carried the fumes away.

The manufacture of phosphorus also produced cases of phossy jaw. At Albright & Wilson's Oldbury works near Birmingham there had been seventeen cases in fifty-four years, but none was fatal. Patients who went down with the disease were given sick-pay of twenty-five shillings a week.

The Salvation Army to the rescue

As we saw in the previous chapter, many in the match industry worked at home making match boxes. A quick worker could, with the help of her children, make as many as forty-five gross of boxes per week (6,480), for which she would receive around 100d (8s 4d), out of which she had to buy paste and brushes. Even so, it could be a useful addition to the household income. Sadly, in many cases this was the only income for a widow and her children and, even when there was a man in the house, he was often able only to find casual employment, in which case match-box making could at least provide a steady source of earnings. The Salvation Army had investigated the match industry, and what they found had appalled them. Their chief investigator was Colonel James Barker, who came across some truly appalling cases, such as the widow with two children under nine who spent sixteen hours a day making match boxes.

Today it is hard to imagine a more selfless group of individ-

uals than the members of the Salvation Army. Back in the
London of the 1880s, however, Salvationists were seen as activists
and trouble-makers. It was during one of his crusades in the East
End that their founder, General Booth, came face to face with
the match girls and learned not only of their poverty but of
phossy jaw. Here was a cause worth crusading for, and he put
forward a plan to eradicate both evils. In 1890 he published his
book *Darkest England and the Way Out*, in which he outlined
his plans to create model factories up and down the land where
people could work and be paid an honest day's pay for an honest
day's work. There would be an end to 'sweating'.

The Bryant & May factory in Bow was one of the better
employers, but the Salvation Army thought it could be a better
employer still. Not only would working conditions in their model
factory be superior but they would solve the problem of phossy
jaw once and for all by making only safety matches:

> 'Lights in Darkest England' – the Salvation Army Social
> Matches!! are now ready and orders can be executed forth-
> with. Everybody should use the 'Darkest England Safeties'
> which are manufactured under Healthy Conditions and are
> Entirely free from the Phosphorus which causes 'Match-
> maker's Leprosy.'

Advertising copy was not a skilled art in Victorian England. This
advert appeared in the Salvation Army's newspaper, the *War Cry*,
in July 1891. But what they lacked in clever phraseology they
made up for in sheer enthusiasm. Here in the world's greatest
and wealthiest city, the capital of the largest Empire that the
world had ever seen, were ordinary people in desperate need –
and what was the Christian message if not to help one another
and relieve human suffering? The plight of those who worked in
the match industry was obvious in the East End of London, the
very part of the city in which the Salvation Army was based.

What the Salvation Army said it needed was a start-up fund

of £100,000 and an annual input of £30,000. As a rallying cry to build a new Jerusalem in England's green and pleasant land, it tapped into a well-tried formula for gaining public sympathy, but as a solution to the social evils of the time it was a non-starter. And yet it started – and started with a model match factory.

The Army bought a disused building in Wick Lane, in the Old Ford district of East London. Commissioner Cadman, the leader of the SA's social section, was put in charge and he appointed George Nunn, a Methodist, as the works manager. Nunn had worked in the industry for many years and had at one time even suffered from phossy jaw himself, but he had avoided the worst ravages of the disease by having all his teeth extracted and by not returning to work until his gums had completely healed over.

The cost of the conversion and of equipping the factory was £2,000, and by the end of March 1891 it was ready to go into production. By mid-April, there were seventy workers employed and on 29 April they registered the trademark of their first brand of safety matches: Lights in Darkest England. The box labels proclaimed General Booth's message of 'security from fire' and 'fair wages for fair work' and were adorned with the Salvation Army crest. Advertisements were placed in the *War Cry*, which carried supportive editorials and ecstatic letters from readers, some of whom sent money:

> Enclosed . . . is a guinea for your first box of matches and a second guinea for a note from the General saying that you have sent me the first box of Darkest England matches. The first ten thousand boxes ought to find ready purchasers at a guinea a box. It just occurs to me that you may have a very handsome offer for the first box, in which case I waive my claim for the good of the social reform funds.

A guinea was 21 shillings. It is clear though that few other readers responded to this exuberant testimony, however genuine they believed it to be. Nevertheless the bandwagon was rolling and, on Monday 11 May, General Booth officially opened the Salvation Amy match factory before a large crowd of well-wishers.

He spoke of it being the first experiment in his Social Scheme and said that it was in the interests not only of the working classes but of the employers as well. He announced that the factory would pay match-box makers 4d per gross compared with the usual fee of 2½d and pointed out that a fast worker, who turned out forty-five gross of boxes a week, could thereby earn 15s, compared with the 9s 4½d paid by other firms. The cost, of course, had to be added to the cost of production, but it worked out at only a third of a penny for twelve boxes.

And there would be plenty of work for them to do. The factory was expecting to produce 2,000 gross of matches a week and they soon hoped to expand the workforce to 100. The General closed his speech by reminding his audience that there would be no deadly white phosphorus used at the factory, and he urged them to support the venture by buying only SA matches and convincing their friends to do likewise.

The first box of matches was then sold for two guineas, the second for one guinea, the third for half a guinea and the fourth for 5s. Other 'numbered' boxes could be bought for 2s 6d a box, and forty-eight people responded, while others bought souvenir boxes at 6d each.

The opening of the factory received national newspaper coverage, with the *Daily News* commenting:

> Whatever opinion may be held with respect to the religious propaganda of the Salvation Army, this is a fact beyond all controversy and when it becomes widely known that there are not big dividends being wrung out of the makers of matchboxes bearing the Salvation Army credentials and that the matches are among the best in the market for money, there will most certainly be a rush for them, unless indeed,

all the sympathy with a most helpless and unhappy class of workers for many years past has been nothing but mawkish hypocrisy.

When the euphoria had died down, the serious business of match-making could begin. Conditions in the factory were clearly better than those in other East End factories, but with working hours from 8 a.m. to 6 p.m., six days a week. This ten-hour day was interspersed with tea breaks at 11 a.m. and 4 p.m. and an hour for lunch, with midday prayers. As an extra treat, there were 'testimony' meetings held every Monday, Wednesday and Friday, which were well attended since most workers at the factory were Salvation Army recruits of the East London Corps.

To begin with, everything went according to plan. All of the Army's publications carried advertisements for the new matches, and the demand was national. A few regional newspapers, such as the *Bridgewater Independent* and the *Spalding Guardian*, offered free advertising space. The new matches were not expensive in absolute terms but they did cost 6d per dozen boxes compared with 1d for lucifers. Nevertheless, orders poured in despite it being mid-summer and, traditionally, the time of least demand for matches. Salvationists were exhorted to visit their neighbourhood shops to obtain orders, and this they did. The *War Cry* spurred on the campaign with poetic tributes such as the one published on 18 September:

It may not be so big a box as you have been receiving;
But what of that, when conscience knocks and tells you it
    was thriving.
The lights you purchased could not be but rank and
    shameful plunder,
The outcome of the slavery, the match-girls laboured under.
. . .
So light no more your 'pipe of peace' with lights that but
    debase it;

But let the sweating match work cease and noble work
    replace it;
And help the people who have planned the future match-
    girls' Eden.
And soon there'll be no more demand for matches made in
    Sweden.

By the end of 1891, the Salvation Army match factory launched a smaller and cheaper version of the Darkest England brand, retailing at 3d per dozen boxes. Even so, they were three times dearer than the price that people normally expected to pay.

Yet, up and down Britain, people were prepared to pay more for their matches as part of the Salvation Army crusade against sweated labour, low wages and phossy jaw. During 1892 the factory in Lamprell Street increased its workforce to 120 to meet the demand, producing six million boxes of matches that year and generating an income of more than £12,500. To increase sales even further, Colonel James Barker decided to concentrate on publicizing the fact that they used no white phosphorus, and he did so by taking a group of journalists and Members of Parliament on a tour of the factory and then on to the home of a family who worked making lucifer matches. To emphasize the point he was making, he would extinguish the lights and, to their horror, the visitors could see the glow of phosphorus on the hands, faces and clothes of the workers.

The crusade against phossy jaw had already borne fruit and, in 1892, special ventilation regulations were introduced governing places where white phosphorus was used and lucifer matches were made. In the Bryant & May factory this had the effect of reducing the number of cases of phossy jaw to fewer than two per year. The regulations were further tightened in 1893. The extra cost of these regulations did not affect the Salvation Army factory, whose success continued. That year they launched two new brands, the Giant Match Tapers and the Boudoir match.

The former were sold in boxes of a hundred and, with extra-long sticks dyed in various colours, they would burn for several minutes and were seen as a substitute for the wax tapers used in places where there was a need to light several things at once. The Boudoir matches were smaller but again made with coloured sticks and were sold in decorous enamelled cardboard boxes that were considered suitable adornment for dressing tables, sideboards and mantelpieces.

The growing success of the Salvation Army matches had some obvious implications for the match-makers. Provided the Salvation Army matches were supplying only a tiny share of the market, they could be tolerated, but it appeared that this share was growing steadily. The time had come to nip the venture in the bud. Shops all over England were suddenly offered Salvation Army safety matches at the same price as ordinary matches. Someone had flooded the market with fakes from Sweden.

The Salvation Army hit back in April by launching the British Match Consumers' League, members of whom pledged to buy only safety matches made in Britain and made in factories where conditions were such that they did not put health at risk. The campaign did little to boost sales, which were now falling badly and threatening the jobs at the factory. In the warehouse, the Army struggled to cope with a mountain of matches that they could not persuade shops to stock, and eventually they had to lay off some of their workforce and put the rest on part-time hours. Adjutant McLauchlan, who was now in charge of the match project, gave an interview to a reporter in which he denied that the firm was on the brink of collapse and pleaded with people to buy their matches. If only all Salvationists would buy them, their troubles would be over.

McLauchlan decided to relaunch their product in July 1894 with better packaging and issued bright-red boxes containing samples of all their matches, which were to be used by agents to generate more business. These trial packs included a new brand, the Briton, which was to be sold at 2d per dozen boxes. These were still twice as expensive as lucifers but were obviously different to the ones being counterfeited. Lists of shops selling only the genuine article were circulated to all Army Corps. Meanwhile McLauchlan decided that the best-selling Darkest England brand should be made slightly larger so that it could be more easily distinguished from the fake boxes, but the price of those was still 6d per dozen boxes.

Each effort to increase sales produced only a temporary boost which quickly fell away again until, by the end of 1894, the factory had to cease production and only a few workers were kept on to service orders from the overstocked warehouse. The Devil had triumphed – but the Angels were not defeated. They would rise again.

McLauchlan analysed the difficulties they were up against in the December issue of the SA's house magazine, the *Officer*. There were three problems to be faced: (1) they manufactured only safety matches, whereas most people preferred lucifers; (2) although they made the cheapest safety matches in England, they still could not compete with some foreign manufacturers of safety matches who were able to sell their matches at the same price as lucifers; and (3) many Salvationists refused to buy SA matches because they were twice and even three times the price of ordinary matches. He quoted one grocer he had contacted who stocked SA matches and who had said that, while he had sold 500 gross of lucifers, a consignment of 30 gross of SA matches was still on his shelves. He even said that among his regular customers were several Salvationists who insisted on buying lucifers. McLauchlan contacted grocers in Manchester, Plymouth and other towns and the story was the same. 'The craze for cheapness over-rules everything else,' was his sad conclusion. The only people who consistently stocked their matches were the Co-ops, and their only consistent purchasers were dedicated trade unionists.

The Salvation Army was reluctant to lose its first venture into solving the problems of Darkest England and decided to run the factory and the sales side of the business separately. The factory became part of the Social Wing, while responsibility for sales were transferred to a Major Laurie at the SA's Trade Headquarters. He decided that they would have more success by targeting institutions rather than individual customers and so he contacted hospitals, mental asylums, prisons, shipping lines, railway companies, religious bodies and the hated workhouses.* Slowly, the stockpile of SA matches was sold, but the factory remained closed for the first six months of 1895. Before it could reopen, more realistic pricing had to be agreed and it was decided to pay 3d per gross to match-box makers, instead of 4d although this was still more than the going rate of 2½d.

By September, new match-making machinery from the US had been installed at the factory along with an automatic drying machine which reduced the time needed to dry the heads of matches to hours instead of days. Together, these made it possible to cut the price of production; alas, it also cut the number of employees, who now totalled seventy, a figure which included home workers.

That autumn the sale of SA matches again began to increase, partly due to direct selling by groups of young boys on the streets of London and partly due to the special offer of a patent match-box holder which sold for 1d. Exports to other countries where the Salvation Army had taken root also accounted for a significant proportion of their output. Despite these efforts, sales refused to rise to a level that made production profitable. Surplus space at the factory was now taken up with making disinfectants, and unprofitable lines, such as the Giant Match Tapers and Boudoir matches, were dropped.

---

* These guaranteed destitute people a bed and food, in return for doing menial tasks.

Ensign Lyne was put in charge of the factory and, in 1897, he installed yet more new machinery which would quadruple output. Now that all the match-making processes were automated, only the box-filling was done by hand. Surely they must succeed. They advertised more widely and took on new sales people. They continued to produce Darkest England matches (now costing 2½d for the smaller size), as well as the Briton brand at 2d per dozen. But, no matter what they did, customer resistance could not be broken down even when the Briton's price was cut to 1½d per dozen boxes in October that year and Darkest England's to 2d. Army newspapers and magazines were full of praise for all these moves, but to no avail.

In any case, a new Act of Parliament now prohibited the employment of young persons in dangerous trades and, while this was also a platform in the Salvation Army's campaign for a better deal for the poor, it ensured that the established match-makers were no longer under the odium of exploiting child labour to make cheap matches. Nor was phossy jaw the threat it had once been.

Still the Salvation Army match factory struggled on but by the end of 1898 even the *War Cry* had ceased to carry advertisements for its products. The final make-or-break campaign was launched in January 1900 with a new advertising campaign which used witty advertisements relating to topical news items of the day. Long features about their model factory appeared in SA magazines, but it was not enough and production languished at only 200 gross of boxes per day. The last advertisement for Salvation Army matches appeared in the 24 February 1900 issue of their magazine *Social Gazette*, still with its proud boasts: of being the only match company in the UK not to use white phosphorus, of employing no sweated labour, of paying higher wages and of having no phossy-jaw cases among its employees. By then, the Bow works of Bryant & May were paying home match-box makers only 1¾d per gross. By a strange irony, after the Army match factory finally closed its doors, it was taken over on 26 November 1901 by Bryant & May.

It is easy to belittle the impact which the Salvation Army had on the use of phosphorus but the days of the lucifer were indeed numbered. The French Government had found a better answer. It had sponsored research into finding a material that could be used to make strike-anywhere matches, and chemists Henri Savène and Emile David Cahen came up with the answer: phosphorus sesquisulphide. Match heads were made with around 13 per cent of this chemical combined with 28 per cent potassium chlorate, the rest being powdered glass, glue and fillers such as iron oxide and zinc oxide.

This curious name, phosphorus sesquisulphide, is no longer used, and today it is called tetraphosphorus trisulphide, the tetra indicating four and the tri indicating three, as shown in its chemical formula P_4S_3. This yellow solid, which melts at 172°C, was made by reacting the two elements in the correct proportions. It has many benefits compared with elemental phosphorus: it is not poisonous, it is not spontaneously flammable, it can be transported easily without the special precautions needed with phosphorus and, above all, it does not cause phossy jaw.

The use of white phosphorus in match-making was finally outlawed as a result of a special international meeting in Switzerland in 1906. The Berne Convention, as it was called, was eventually signed by all nations except the US and it obliged signatories to enact laws prohibiting the manufacture of white phosphorus matches. The British Parliament passed a law in 1908 that made them illegal after 31 December 1910 and in so doing it joined the ranks of countries which had long banned the use of white phosphorus in matches.*

---

* Finland had outlawed them as long ago as 1872, Denmark in 1874, Sweden in 1879, Switzerland in 1881 and Holland in 1901. Some countries resisted implementing the international agreement for several years, and

The US said it could not sign the Berne agreement on constitutional grounds and instead imposed punitive taxes on phosphorus matches in 1913 which ended their manufacture. But, in any case, US factories were much cleaner and better run, and match-makers even claimed that they had never encountered cases of phossy jaw. This turned out not to be true and, when enquiries were made and reported on in 1910, some 150 cases came to light. Even as late as 1928 Dr E. F. Ward, in the *Journal of Industrial Hygiene*, reported fourteen cases of phossy jaw in US firework factories, of whom two died of the disease.

Phossy jaw and phosphorus matches did not entirely disappear as a result of the Berne Convention. As late as 1950 the *Tientsin Daily* reported that they were still being produced in parts of China. There, the former Kuomintang regime had permitted many local match works to continue making lucifers and even to stay in business after the Communist revolution, with the result that six workers had been treated for phossy jaw and that one of them, Chang Chia Hsiang, had died of the disease.

Phosphorus and phossy jaw

We will never know for certain why phosphorus caused the disease phossy jaw. Attempts earlier this century to induce the disease in rats and study it that way came to no conclusion. A few cases of phossy jaw continued to occur in factories making phosphorus for military purposes, although in the UK, at locations where phosphorus incendiary bombs were made on a large scale in the Second World War there were no cases. In 1944, R.

India and Japan only banned them in 1919, while it took pressure from the League of Nations, forerunner of today's United Nations, before China agreed to ban them in 1925.

Kennon and J. W. Hallam reported on seven men with phossy jaw in an article in the *British Dental Journal* and, in 1960, there was a paper in the *British Journal of Industrial Medicine* which described ten cases.

In 1946, H. Heimann of the New York State Department of Labor reported on three cases that had occurred in a plant making elemental phosphorus. What is puzzling is that the condition was not necessarily caused by poor dental hygiene and it seems it could develop even when workers were given regular check-ups, including dental X-rays. Heimann supported the theory that elemental phosphorus was absorbed by the lungs and thereby entered the bloodstream, weakening the whole skeleton and not just the jawbone. Evidence for such a systemic cause was the known susceptibility of workers in the phosphorus industries to spontaneous bone fractures.

It was even harder for Victorian doctors to explain phossy jaw and they naturally associated the condition with dental caries. Moreover, they were still of the opinion that phosphorus itself was a medicament. Dr Cunningham, one of the commissioners who wrote the 1892 report, thought that working with phosphorus could be good for the health. In his words:

It must not be imagined that phosphorus is nothing but a poison. It is sometimes prescribed ... as a nerve-tonic. It is therefore not impossible that, under favourable conditions and where very little was taken into the system, it might be beneficial. Whence it may follow that in a well-conducted factory, with every precaution taken and plenty of ventilation, no ill results to health might follow. In fact, there is reliable evidence that apart from necrosis the general condition of health of match-makers is not unsatisfactory. ... the general health of operatives exposed to phosphorus fumes compares very favourably with that of those employed in other factories under similar conditions in the same district.

The 'reliable evidence' to back up this last statement was a few statistics from a paper by a Parisian physician, Dr Mahu, who had observed that in the years 1891 to 1898 there had been thirty deaths from TB in a local tobacco factory but only twenty in a nearby match factory and concluded that death from tuberculosis was considerably less in the latter occupation compared with the former. Cunningham found this to be convincing evidence that phosphorus had some preventative power.

Dental caries appeared to be the main way in which the phosphorus gained access to the pulp cavity of the tooth, thereafter working its way into the jaw. Bad teeth were a simple fact of life for most working people of that era. Cunningham quoted figures provided by firms that made and used phosphorus. People in these industries who had teeth that were described as 'bad' or 'very bad' amounted to 697 people out of a total of 965 examined: that is, 72 per cent.

Cunningham was certain that most of the damage the patient suffered was due to infection by micro-organisms. Experiments had shown that the pus that discharged from phossy jaw contained staphylococcus, streptococcus and other bacteria. But how was phosphorus able to bring about such a terrible condition? One theory is that fumes of phosphorus in the air led to the formation of phosphorus oxide, which was absorbed by the moisture in the nose and mouth thereby forming phosphorus acids, which then attacked the teeth. The cavities thus formed then became progressively worse as the acid ate its way through the enamel and down the pulp cavity until it attacked the jawbone itself, aided and abetted by bacteria.

All we can say for certain about phossy jaw is that the most likely route whereby phosphorus entered the body was absorption from the air via the lungs and into the bloodstream. This was then capable of affecting the skeleton, the jaw in particular, causing a weakening of the bone and pain. If the victim's teeth were already in poor condition, with infection reaching down to the roots, the combination could produce the terrible symptoms associated with the disease.

It is difficult in today's carefully regulated working conditions, with public awareness of health and safety issues, to understand the general acceptance of phossy jaw and the seeming reluctance to stamp it out. Even the vigorous publicity campaign aimed at raising public awareness by the Salvation Army failed to lead to effective government action beyond an investigative commission's recommendations. Instead the SA directed its efforts into trying to provide a practical answer to a social evil, but this course of action will never succeed against an established industry. Nevertheless it was a brave attempt. The only way to stamp out phossy jaw was to eliminate white phosphorus, something that could have been achieved half a century before, if only . . .

7. Gomorrah

Playwrights and novelists have described the dramatic effects of burning phosphorus to portray the horrors that civilians faced in the Second World War (1939–45), when attempts were made to destroy whole cities with it.

> HEISENBERG: You never had the slightest conception of what happens when bombs are dropped on cities. Even conventional bombs. None of you ever experienced it. Not a single one of you. I walked back from the centre of Berlin to the suburbs one night, after one of the big raids. No transport moving of course. The whole city on fire. Even the puddles in the streets are burning. They're puddles of molten phosphorus. It gets on your shoes like some incandescent dog-muck – I have to keep scraping it off – as if the streets had been fouled by the hounds of hell. It would have made you laugh, my shoes kept bursting into flames. All around me, I suppose, there are people trapped, people in various stages of burning to death.

That is taken from Act I of *Copenhagen* by Michael Frayn, a play which centres round the meeting in 1941 between Niels Bohr, the Danish physicist who developed atomic theory and who helped the Americans develop the atomic bomb, and Werner Heisenberg, the German physicist who was developing atomic power for the Nazis but who held back from making a bomb.

In his book *Bomber*, Len Deighton describes the effects of a phosphorus bomb:

'Bombs gone,' said Digby, his voice completely relaxed.

When the two 250-lb phosphorus bombs from Creaking Door [the name of Digby's bomber] hit the old building the quick dropped flat upon the dead. The phosphorus bombs threw their showers of white sparks sixty feet into the air and they came down as pretty as fireworks with large fragments of burning phosphorus. A cluster of fragments struck Johannes Ilfa's legs. He knifed it away from the fabric and flesh but some were deeply embedded. Around him men screamed and fought and burned and sometimes survived. They wriggled like live bait and a few regained their feet like ghosts arising from a mass grave. Water had no effect upon phosphorus. The official instructions were to cover it in sand. Can you put sand upon a man's face?

Perhaps it is impossible to exaggerate the horror of what phosphorus bombs can do, but other writers have tried to capture the moments of sheer terror:

> But the worst was yet to come . . . a large phosphorus bomb fell directly outside . . . the people nearest the door now gave way to an indescribable panic . . . Terrible scenes took place, since all of us saw certain death in front of us, with the only way out a sea of flames. We were caught like rats in a trap. Doors were thrown on the canister by screaming people and more smoke and heat poured in . . . some collapsed and never woke up again. Three soldiers committed suicide. I begged my husband to beat back the flames with our blanket but he was unable to do so. My hair began to singe . . .

What distinguishes this account from the first two is that it is not from a play or novel but is the actual account of a woman who miraculously survived a phosphorus bomb. This fell near the entrance of the public toilets where she and her husband had taken shelter and a shower of molten phosphorus cascaded in upon them. Their story is to be found in tales of the bombing of

Hamburg, and the above account is taken from the report of the raid compiled by the office of the Chief of Police in Hamburg. It is quoted in Martin Middlebrook's book, *The Battle of Hamburg*, in which he recounts the experiences of those who survived the deluge of phosphorus bombs that were dropped on that city in an attempt to obliterate it in a week of air raids in 1943.

△

It only became practicable to wage war with phosphorus when it could be produced in quantity. Large-scale production became possible in 1882, when electric furnaces were introduced which enabled phosphorus to be extracted from phosphate ore in a continuous process. The demand for phosphorus was to receive a boost in the First World War, when it was mainly used to create smokescreens to hide troop movements. Production in Britain increased from 1,000 tons in 1914 to 2,500 tons in 1918. Burning phosphorus creates clouds of dense white smoke that make an ideal screen, and so it became an essential part of military attacks on the Western Front. Care had to be taken to fire the shells well ahead of the advancing troops in order to avoid spraying them with fragments of molten phosphorus, which could adhere to clothing and skin and inflict terrible wounds. For this reason, a barrage of such shells might well be aimed directly at the enemy, causing them to flee in terror.

Another use of phosphorus that had both defensive and offensive possibilities was tracer bullets, which relied on the burning phosphorus to mark out their trajectory at night, thus enabling a gunner to direct his fire more accurately. There were also Mauser explosive bullets containing one-twentieth of a grain of phosphorus (*ca* 3 g) and these could also deliver a lethal blow to a highly flammable target such as a Zeppelin.

When these hydrogen-filled airships were sent to bomb London in 1915, they did so with impunity because they were able to fly at heights well above fighter aircraft. Zeppelins were held aloft by 33,000 cubic metres of hydrogen gas (1.1 million cubic feet), all of which was potentially highly flammable – if only it could be ignited. To begin with, the advantage was with the Zeppelins; they could travel at 100 k.p.h. (60 m.p.h.) and, although their bomb load was small, they could find their target easily because of the lights in central London. The panic they caused was out of all proportion to the threat they posed to life and property.

Eventually, the British built fighters that could attack them and equipped these faster planes with special .303-inch incendiary bullets which ignited a charge of phosphorus as they were fired. As a result there were some spectacular nights when Zeppelin bombers exploded over southern England. The Germans tried to counter this threat by building lightweight Zeppelins that could climb even higher, and, with their undersides painted black, they were difficult to see against the night sky and hard for searchlights to find, but the war ended before they could be used.

Phosphorus really came into its own as an offensive weapon in the Second World War. Smoke bombs and shells were still to be important and were now filled with a mixture of phosphorus and a rubbery gel. When such a shell exploded, the phosphorus was more finely dispersed, thus creating a more effective smokescreen. Phosphorus might even be used by civilians in defence of their homes.

In May 1940, the German Army invaded the Netherlands, Belgium and France and within weeks all had capitulated. The British Army retreated to the French port of Dunkirk and was evacuated across the English Channel, leaving all weaponry behind. Hitler then began the Battle of Britain in an attempt to defeat the RAF as a prelude to invasion. The British Government ordered the production of millions of Molotov cocktails, named after the then Soviet Foreign Minister. These were simple incendiary devices that civilians could use against the invaders. They were bottles filled with a solution of phosphorus dissolved in benzene and, when they were thrown, would smash and the contents immediately catch fire.

The main phosphorus manufacturers, Albright & Wilson, were asked to make 250,000 of these a week, using whatever was to hand. They commandeered the production of screw-top beer and milk bottles, creating a national shortage of these items, and, by June 1940, they were producing these simple weapons in an emergency factory in Kidderminster, employing the female workers from a local carpet mill to work the bottle-filling machines. The completed Molotov cocktails were distributed around the country and stored in crates in local streams for safety. During the following two years more than seven million were made, but they were never used for their intended purpose. Many ended up floating away down rivers, and crates of them continued to turn up for many years after the war.

The production of Molotov cocktails, though, was only a minor diversion. The scattering of phosphorus bombs over cities was to inflict real damage. The most common type of incendiary bomb used burning magnesium to start a fire. These were 4 lb (1.8 kg) thermite bombs which relied on the chemical reaction of aluminium and iron oxide to generate enough heat to ignite the magnesium casing of the bomb. Once this was alight, it would burn fiercely, but such bombs, which were about as big as two modern drink cans, were often too light to penetrate roofs and, even if they did, they could be scooped up and thrown into the street.

Phosphorus bombs, on the other hand, were much heavier, generally weighing about 30 lb (14.4 kg), and these were designed to have sufficient momentum to penetrate deep into a building, where a small explosive charge would scatter burning phosphorus all around, quickly creating a raging fire as the molten element ran everywhere. If the inhabitants of the building were sheltering in the basement, they would have to get out quickly. When such bombs fell in the street, they presented a terrible hazard for the defenders and fire-fighters, as well as for fleeing civilians.

The instructions issued to US civilians at the time indicate the nature of the threat:

Phosphorus bombs: ... A 30 pound bomb may scatter burning particles over a considerable area. The phosphorus ignites upon contact with the air and requires air in order to burn. Water may be used to extinguish the phosphorus but as soon as it dries off it will reignite. The phosphorus particles must be collected and taken to some location where they may be buried or allowed to burn out. All of the area in the vicinity of such a bomb must be examined to locate these particles. Burning phosphorus gives off a characteristic white smoke which is readily detected. Sand or water form good extinguishing agents. A solution of copper sulphate forms a coating on the particles and prevents it from reigniting in the air. Burns from phosphorus are severe and continue until the particles are removed from the skin or flesh. A 2% solution of copper sulphate may be applied to the skin which stops the burning of the particles and they may be removed. Do not handle any of the particles with the bare hands.

US civilians were never exposed to air raids, although, as we shall see, some Japanese phosphorus bombs did reach North America. In Europe the bombing of cities became a nightly event for five long years. The Nazis were the first to resort to this strategy as a way of destroying industry and demoralizing citizens. For nine months, from September 1940 to May 1941, the bombers of the Luftwaffe sought out British cities and bombed them night after night. Their favourite target was London, then the largest city on Earth and it was vast, covering more than 750 square kilometres (300 square miles), and with eight million inhabitants. It was particularly vulnerable because it was only

200 kilometres (120 miles) from Nazi-occupied territory and suffered frequent air raids in which the Luftwaffe dropped mainly high explosive and magnesium incendiaries. London could absorb such punishment because of its size, and there was an army of firefighters to deal with burning buildings, and a legion of workmen to clear the debris and repair the damage.

Nine months of nightly bombing failed to bring the Nazis the victory they sought in the west as a prelude to attacking Russia, but Hitler could wait no longer. In June 1941, his forces moved east and air raids on British cities were scaled down as most bombers were moved to the Russian front. The British had not been beaten by heavy bombing, but while they suffered they planned their revenge. At the time it was said the Nazis had sown the wind and soon they would reap the whirlwind – but it was not that easy.

When the RAF tried to bomb German cities, they generally missed them because they lacked the sophisticated equipment that had guided German bombers to towns and cities in Britain. An analysis of the flash photographs that bombers took when they dropped their bombs showed that most fell on open countryside. When 4,000 such photographs were analysed by a civil servant, D. M. Butt, in November 1941, he concluded that only one bomber in four got to within *five* miles (8 kilometres) of the intended city in most raids and that many were off target by as much as *fifty* miles (80 kilometres). In some raids, only one bomber in ten found the right location.

In the face of such evidence, the bombing of Germany was suspended while better navigational equipment was devised and a more effective method of attack was introduced. Within a few months the RAF had both radar detectors that could scan the ground and pinpoint towns and cities, and a way of marking the target with coloured flares dropped by pathfinder aircraft. A stream of bombers could then deliver a concentrated blow in what was termed 'area bombing'. No attempt was made to identify industrial or military targets; the object was simply to obliterate built-up areas and to do it quickly so that the defences were overwhelmed.

The pride of RAF Bomber Command was the Lancaster bomber, which generally carried a mixed load of one 4,000 lb high-explosive bomb, 960 of the 4 lb magnesium bombs and 64 of the 30 lb phosphorus bombs. The blast from the high-explosive bomb was designed to blow off roofs and blow in doors and windows, so that the incendiaries which dropped into buildings would have plenty of air, thereby ensuring fires took hold rapidly.

The success of the new bombing campaign came in the spring of 1942 when two Baltic ports, Rostock and Lübeck, were destroyed in heavy raids. The Nazis retaliated by bombing historic cities in Britain, namely Exeter, Bath, Norwich, Canterbury and York, all of which got a three-star rating in the famous German tourist guide, *Baedeker*. These so-called Baedeker revenge raids could only be carried out by a relatively small force of bombers with the result that damage was serious but limited mainly to historic buildings, shops and houses, and casualties were relatively few.

On the other hand, by the summer of 1942, the RAF had a fleet of 1,000 bombers and the decision was taken to launch all of these against a single target. The raid on Cologne on the night of 30 May 1942 was the first attempt by the RAF to destroy completely a large city using two-ton and one-ton high-explosive bombs, 450,000 magnesium incendiaries and 7,500 phosphorus bombs. However, the majority of these failed to fall on Cologne, despite RAF pathfinders locating and marking the centre of the city with flares.

Bombers tended to unload their bombs at the first sight of fires on the ground, with the result that they often bombed decoy fire sites, of which there were many surrounding the major cities. Those which did bomb the actual target tended to unload their bombs as early as possible, so that a 'creep-back' developed and soon bombs were falling on the suburbs and then on the surrounding countryside. However, the raid on Cologne was moderately successful.

The bombs which actually fell on the city were recorded by the office of the Police President of Cologne, which estimated that these consisted of 20 two-ton and 800 one-ton high-explosive bombs, 110,000 magnesium incendiary and 560 phosphorus incendiary bombs. If these estimates were correct, then around a quarter of the attacking bombers found their target. They hit 3,300 residential buildings, completely destroying 13,000 houses and apartments and damaging a further 30,000 homes. Thirty-six factories were completely wiped out and seventy badly damaged. Around 1,000 commercial premises were destroyed and about the same number damaged. In human terms the totals were: 470 people killed, 5,000 injured and 45,000 rendered homeless. The damage was serious, but when the RAF attempted to repeat their success, with a 1,000-bomber raid on Essen the following night, almost no bombs fell on the city. Their aim was only a little more accurate when they sent 1,000 bombers against Bremen three weeks later, although in that raid an important aircraft factory was hit.

During the following year other German cities were bombed with some notable successes, but, if bombing were to be the road to victory, the RAF still had to prove that it could deliver a crippling blow. And so, in July 1943, Bomber Command decided to show the people of Nazi Germany what fate was in store for them: they would obliterate Hamburg, the Reich's second largest city and Germany's largest seaport, with 1,750,000 inhabitants.

Operation Gomorrah

There were good tactical reasons for choosing Hamburg: it was a centre of U-boat construction – of which it made 408 during the war years – and these were a major threat to allied shipping crossing the Atlantic. It was also a key manufacturing centre and storage area, but it was large, covering 130 square kilometres (50 square miles) and, while this meant it was easy to hit, it would be difficult to destroy.

By July 1943, there had already been ninety-eight air raids on Hamburg, but these had achieved very little in terms of damage and most had been merely 'nuisance' raids, the latest of which had been by three Mosquito bombers early in July. Hamburg had originally been intended for the first 1,000-bomber raid, which would have occurred the previous summer and exactly one hundred years since the Great Fire of Hamburg in 1842, but the weather forecast for the city was of thunder clouds and so Cologne had been chosen instead. Although it had been given a year's grace, Hamburg's time was now up.

The destruction of Hamburg had been planned for some time, and how it was to be achieved was outlined in the top-secret Operations Order No. 173 dated 27 May 1943. This estimated that 10,000 tons of bombs would be needed, the majority of which would be incendiary bombs, half phosphorus bombs and half magnesium. A high level of phosphorus bombs was chosen because of the well-known effect they had on morale.

The attack, code-named Operation Gomorrah, was scheduled for the last week in July 1943. Five major night-time raids were planned for the nights of 24, 25, 27, 29 July and 2 August, with day-time raids by the United States Air Force on 25 and 26 July. Interspersed with these seven major blows would be lighter day-time raids so that the defenders of the city got no rest. The Americans now had 300 B-17 bombers (Flying Fortresses) based in England and were willing to take part, provided they engaged in precision bombing of key sites, which was possible in daylight hours by these heavily armed aircraft. At this stage in the war the idea of area bombing was anathema to the USAF, whose pilots baulked at the idea of dropping bombs on defenceless women and children.*

Although the Hamburg Fire Brigade was superbly equipped with more than 300 modern fire-tenders, 50 fire-ships, 900 auxiliary pumps stationed at sensitive sites and 3,600 men, they could not cope with what was about to happen. Nor could those who were charged with protecting the civilian population. When the Germans had realized, early in 1942, that they were to be subjected to a prolonged bombing campaign, they had begun to construct bomb-proof air-raid shelters for people living in the inner cities, and by July 1943 such shelters in Hamburg were capable of holding 378,000 people. The vast majority of the citizens, however, had only the reinforced cellars of their apartment blocks in which to shelter from the bombs.

Day 1 – Sunday 25 July 1943. That summer Hamburg had basked in a heatwave and, for two weeks, the day-time temperatures had approached 30°C, while little rain had fallen. By the last week in July the city was tinder-dry.

The first bombs of Operation Gomorrah fell on Hamburg at 1 a.m. on Sunday morning, 25 July. By 3 a.m., when the all-clear sounded, 1,400 tons of high explosive and 1,000 tons of incendiaries had descended on the western part of the city, large parts of which were in flames, including the infamous red-light district of St Pauli with its notorious Reeperbahn.

The incendiaries consisted of 27,000 of the 30 lb phosphorus bombs and 330,000 of the 4 lb magnesium bombs. Some of the phosphorus bombs went off spectacularly as they dropped, creating large fireballs and falling in streams of molten phosphorus, cascading down on to the roofs and walls of houses. After the raid, there was much talk about the liquid which dropped in

---

* Later in the war, though, in 1945, they did engage in area bombing against Japanese cities.

glowing streams and people wondered what it was. So began the myth that the RAF had subjected Hamburg to a 'rain of phosphorus', although in fact many witnesses had mistaken the target indicators for phosphorus bombs. The indicators descended slowly and resembled cascades of fire as they fell.

In some districts the concentration of fire bombs was sufficient to cause huge conflagrations. St Pauli, Altona and Eimsbüttel were all badly hit and more than 1,200 people were killed in the raid. The Hagenbeck Tierpark found itself beneath the rain of bombs and its world-famous zoo was almost completely destroyed, along with 140 of its rare creatures. Those which survived, including most of the elephants, were rounded up and herded into wagons to be rehoused at Vienna zoo.

One resident, Frau Rosa Todt, recounted what happened to people who were splashed with phosphorus:

> they presented a fearful sight. Their skin was bright red, water dripping out of the pores of their skin; their ears and nose, their whole face, was a nauseating mask.

During that Sunday afternoon, the USAF mounted a daylight raid by B-17 bombers. These Flying Fortresses carried either ten 500 lb high-explosive bombs or sixteen 250 lb phosphorus incendiaries, and 123 of them set off to attack Hamburg. Ninety bombers reached the city, where about half of them delivered their load on to the designated targets, the shipbuilding yards and those factories vital to the war effort. The plan worked well and the Howaldtswerke yards, the MAN diesel-engine works and several oil-storage tanks were hit. Other parts of Hamburg were also blasted, including the train loaded with zoo animals which was standing near the main station in the city centre.

That night, the RAF had intended to attack the city again, but the bombers were diverted to Essen instead because the pall of smoke over Hamburg was such that target-finders thought it would be impossible to place aiming flares accurately enough.

Even so, Hamburg was not allowed to sleep and a group of six Mosquitoes attacked the city, dropping four and a half tons of bombs.

Day 2 – Monday 26 July 1943. On the morning of Monday 26 July the USAF again bombed Hamburg but, of eighty-two Fortresses dispatched, twenty-five turned back due to technical difficulties and only fifty-three attacked the city, dropping 91 tons of explosives and 27 tons of incendiaries. Again they targeted the docks area, especially the U-boat shipbuilding yards of Blohm & Voss, which were badly hit, as was an aircraft-assembly shop, two floating docks and two large gasholders. Their biggest success was in putting the Neuhof power station, which was Hamburg's largest supplier of electricity, completely out of action. This was achieved by one 500 lb bomb that destroyed the all-important boiler house, thereby doing more economic damage to the city's industrial output than all the other bombs dropped that day. Chiefly as a result of this raid, U-boat production dropped from a planned fifty-eight to forty-three for the months of August and September.

That evening there was another nuisance attack by six Mosquitoes.

Day 3 – Tuesday 27 July 1943. Tuesday started hot and, as the day wore on, the temperature soared. Hamburg continued to swelter under a summer heatwave, but at least its remaining inhabitants were able to go about their daily business without the disruption of another air raid. The USAF bombers stayed away.

Meanwhile, the RAF was preparing for another major attack on the city. At 22.00 hours a fleet of 787 bombers began to take off from airfields in England and, of these, 738 would reach the city, delivering more than 3,000 tons of bombs, of which the vast majority fell in the target area. This time more incendiary bombs were to be dropped and the bombing was scheduled to start at 1 a.m.

Die Katastrophe was about to begin.

Target-markers dropped coloured flares over the central area of Altstadt and the bombers approached from the east of the city, dropping their bomb loads on the mainly working-class districts of Hammerbrook, Borgfelde, Hamm Süd and Hamm Nord. It quickly became apparent to those on the ground that this was again to be a heavy raid. Within fifteen minutes major fires were raging in some areas and these were combining to form conflagrations. Bomber crews were arriving to find a sea of flame below them. The first hint that something strange was happening was at around 1.30 a.m., when fires began raging out of control, with one building igniting the next. Yet, for another half-hour, more bombers were to drop more phosphorus on to the inferno. And then the firestorm started.

Rather special atmospheric conditions are required to create a firestorm, and these existed that night in Hamburg. The tens of thousands of citizens in the affected areas, who were cowering in their cellars, had only about thirty minutes after the raid had ended to leave their shelters and escape to open ground. By 3.30 a.m., the firestorm was raging with hurricane-force winds, and anyone who ventured into the street was sucked into the inferno. In less than an hour, the carpet of burning magnesium and phosphorus bombs that had been laid over this part of the city had set enough buildings on fire to make fire-fighting impossible and, as their roofs collapsed and windows spewed forth great jets of flame, the rising heat began to draw in air along the surrounding streets.

In his book *Operation Gomorrah*, Gordon Musgrove describes the plight of Herr Wilkan and his wife Erica, who lived in an apartment at 83 Grevenweg, Hamm Süd, near the Mittelkanal. When the alarm sounded they joined the other residents in the cellar, but, within thirty minutes of the raid starting, their apartment block was on fire and smoke from the burning building forced them to leave. Outside, they discovered that the whole street was ablaze, so they ran towards the canal where they knew they could shelter in a large public toilet block. There they were soon joined by a hundred others who had had the same idea. It was their experience which was recounted, in part, at the start of this chapter.

As the raid continued and the air became hotter, they used the water from the cisterns to wet their faces and clothes to try to keep cool. Then an incendiary bomb exploded near the entrance and those near the door were covered with burning molten phosphorus. Those at the rear took the cubicle doors off their hinges and used them to prevent the molten phosphorus from running into the toilets, but it was to no avail. People's clothing caught fire and they sank dying to the floor, despite the attempts of those with wet blankets to douse the flames. Three young soldiers who had been drenched with droplets of burning phosphorus put an end to their agony by shooting themselves.

The Wilkans decided that they would have to leave and, along with about twenty survivors, they wrapped wet blankets around their bodies and ran out of the toilet block towards the canal. There they rested and waited for the firestorm to abate. They had survived and, later that morning, they trudged down the autobahn to Barsbüttel, where they were taken care of by a local family.

Why a firestorm should have developed that night has been much debated. As the cities of Germany succumbed to mass bombing, only a few of them, such as Wuppertal, Krefeld and Dresden, experienced the horrors of a firestorm. Firestorms were not new, although historically they were rare. When Rome burned out of control in the time of Nero in AD 64 and London in 1666, no firestorms developed, and yet, when Hamburg had previously burned in 1842 and Chicago in 1871, they did. The Great Fire of Hamburg started on 5 May 1842 in a small merchant's property in the Deichstrasse, near the waterfront of the River Elbe. It spread quickly and burned for three days, by which point the heart of the city had been destroyed.

Rather ironically, the key factor deciding a burning city's fate is the *absence* of wind. The rising heat from the fires has to remain concentrated and, as it starts to rise, it creates the equivalent of a stove-pipe or tall chimney in the atmosphere. The hot air rises up as though in a flue, which in turn sucks in air at ground level directly into the area of the conflagration, thus making the fires burn more fiercely and enabling jets of flame from burning buildings to ignite all around.*

Returning bomber crews reported a two-kilometre-wide column of smoke rising above Hamburg, and this would eventually reach a height of five kilometres. Anyone caught up in the thirty square kilometres of blazing furnace feeding this chimney would have stood no chance of survival, whether they stayed in shelters or took to the streets. Only those residents who had left their shelters immediately after the raid finished and moved quickly out of the area would have lived. But for those who delayed doing so, the firestorm would have sealed their fate. By five o'clock, it was subsiding and, by six, it was almost over, but an estimated 30,000 men, women, children and babies were dead. When the authorities later tried to count the survivors among those who had been residents in Hammerbrook, one of the worst-affected districts, they could trace only a hundred of them.

Operation Gomorrah was living up to its name – but the

---

* Somewhat surprisingly the destruction of Sodom and Gomorrah is described in the Bible (Genesis 19:28) in similar terms, suggesting a firestorm also devoured those twin cities:

And Abraham went early in the morning to the place where he had stood before the Lord: and he looked down toward Sodom and Gomorrah, and toward all the land of the valley, and beheld the smoke of the land went up like the smoke of a furnace.

programme of bombing was only half complete. Two more nights of burning phosphorus were planned. By now, a million people were fleeing the city, as it had become apparent that the RAF was out to destroy it.

Day 4 – Wednesday 28 July 1943. The smoke from the previous night's bombing ruled out any day-time raids by the USAF or night-time raids by the RAF. However, there was a nuisance attack by four Mosquitoes, designed to sound the air-raid sirens and so rob the city's remaining residents and defenders of another night's sleep, while the crews of Bomber Command rested in their beds.

Day 5 – Thursday 29 July 1943. Although it may seem impossible, the third raid by the RAF was to inflict even more material damage to the city than had been achieved on the fateful Tuesday night. By the time this raid was over, the sprawling north-west suburb of Barmbek would lie in ruins.

Reconnaissance aircraft reported that the weather over Hamburg was clear and, although fires were still burning from the Tuesday raid, the smoke was not likely to interfere with pathfinders locating the city and dropping target-markers for another heavy raid. The focalpoint was the same, the Altstadt in the centre of the city, but this time the approach would be from the north, so that the bombs could be dropped on the so far unaffected middle-class suburbs of Rotherbaum, Harvestehude, Hoheluft and Eppendorf, which straddled the Alster, the large lake in the centre of Hamburg. The attack would begin at forty-five minutes past midnight.

In fact, a prevailing wind was now blowing across the city from the south-west and this carried the path-finder aircraft to the east, so that the markers were dropped in east Hamburg just to the south of the firestorm area. Consequently, the oncoming bomber stream began dropping its loads on the still smouldering

Hamm, but this carpet of bombs soon crept northwards so that most landed on the Barmbek district. Within a short time this too had become a massive conflagration as many fires raged out of control and several important factories were destroyed.

In terms of human lives lost, the third raid was relatively light and only 800 people perished, half of whom died in one of the shelters of the Karstadt department store on Hamburgerstrasse. A high-explosive bomb brought down part of the building and blocked the exits to the shelters. Although these were unblocked the following day, releasing 1,200 unharmed from one of the shelters, those in the other shelter were all dead, killed by carbon-monoxide poisoning from a nearby coke cellar which had been ignited by an incendiary bomb.

Day 6 – Friday 30 July 1943. Another massive raid was planned but was shelved by a sudden decision to bomb towns in northern Italy instead. The war was going badly for Italy, the allies had successfully invaded Sicily and the Fascist Grand Council finally voted Mussolini out of office – and imprisoned him. It was thought that heavy bombing raids on Milan, Turin and Genoa might convince the Italians to sue for peace, thereby sparing their cities the fate that was now befalling those of Germany. These raids were called off at the last minute but too late to allow Bomber Command to reinstate the planned raid on Hamburg.

Days 7 and 8 – Saturday 31 July and Sunday 1 August 1943. These were a wash-out, literally. Two bad-weather fronts passed over the British Isles and moved across the North Sea to Germany. Preparations for another massive raid on Hamburg had just been completed on the Saturday but then there was a series of violent thunderstorms across eastern England, where the bomber airfields were situated, and so the operation was cancelled. The bad weather also meant that the following night's raid was cancelled because it made flying dangerous both across the North Sea and in northern Germany. Even Mosquito nuisance raids were cancelled.

Day 8 – Monday 2 August 1943. To the south of Hamburg is Harburg. This had so far escaped attack, but it also harboured key industries. Seven hundred and forty bombers were dispatched to take its factories out of the war effort, and again the weight of incendiaries carried was larger than that of high explosives.

However, the weather was still bad and 186 bombers turned back because of icing and turbulence over the North Sea. Others, blown off course, sought alternative targets, dropping their bombs on Bremen, Cuxhaven, Wilhelmshaven and Bremerhaven before heading for home. About 400 bombers reached the Hamburg region, only to find it completely covered in cloud; target-indicators could not be placed accurately and so those planes which dropped their loads did so blind. Many of the bombs fell on villages around Hamburg, to the extent that many believed it was part of a plan to bomb refugees from the city. The Hamburg defenders estimated that 300 bombers hit the city itself that night and, for the remaining inhabitants, it was the worst raid yet because it went on all night, bombs fell everywhere and a violent thunderstorm raged overhead. The City Opera House was set alight but prompt action by fire crews saved it from complete destruction. By now, no one was keeping a record of casualties, but there were probably very few.

In terms of damage caused, the fifth and final raid of Operation Gomorrah was a failure but, at least for the inhabitants, the Battle of Hamburg was over. Those still in the city expected another big raid because many districts, including industrial ones, were largely intact, but the port was left to recover and was not seriously bombed again until the following summer.

Although the main weight of bombs dropped consisted of high explosives (7,000 tons) and magnesium incendiaries (1.5 million), the raiders also dropped 1,900 tons of burning phosphorus on to the city in bombs of various sizes ranging from 20 lb (9 kilograms) to 250 lb (90 kilograms). However, while it was possible for those defending their homes to deal with a magnesium bomb, there was no defence against a direct hit by a phosphorus one.

At the end of that week 25 square kilometres (10 square miles) of the city were reduced to rubble and 800,000 people were homeless. At least they were alive. The total number of civilians killed in the Hamburg raids was 37,000, but 10,000 were missing, presumed dead. This can be compared with the 30,000 killed by bombs in London during the years 1940–5 but falls far short of the 80,000 who died in an incendiary raid on Tokyo by the USAF on the night of 9 March 1945; or the 140,000 who died when an atomic bomb was dropped on Hiroshima on 6 August that year; or the 70,000 people killed when another fell on Nagasaki three days later.

In Hamburg most of the damage, deaths and disappearances happened during the firestorm on the night of 27 July. When some of the blocked air-raid shelters were eventually reached after the raid, the corpses were so badly burned that only rough estimates of the number of bodies could be given.

The rest of Hamburg slowly recovered and, by the end of 1943, the city was functioning again, 270 kilometres (170 miles) of rubble-filled streets had been cleared for traffic and production of material for the war effort was 80 per cent of the pre-raid level, thanks to workers being drafted into essential industries from those factories which had been destroyed. The city's newspaper, *Die Hamburger Fremdenblatt*, was on the streets again by 18 August, filled with uplifting stories of how well the citizens were coping. Since large areas of Hamburg were now off-limits to the population – access to the so-called Dead Zone being

---

Operation Gomorrah: the reckoning

People killed at least 37,000

Property destroyed

253,000 apartment blocks	77 cinemas and theatres
35,500 houses	76 municipal buildings
2,632 shops	69 post offices
580 factories	58 churches
379 office blocks	24 hospitals
277 schools	1 zoo
83 banks	

Water mains broken 850

Shipping sunk 180,000 tons

Rubble 40 million tons

---

completely forbidden without a special pass – it was difficult for the ordinary citizen to estimate the scale of the disaster.

Half the apartments in the city had been destroyed, but the problem of accommodating people bombed out of their homes was not as pressing as might have been expected. More than half of the population had fled, although 200,000 were to drift back by the end of the year. The city's transport system was paralysed for months; electricity and gas services were severely disrupted. Operation Gomorrah was incredibly destructive (see box) but it fell far short of the knock-out blow that might have convinced Hitler that the time had come to sue for peace.

Operation Gomorrah did affect the outcome of the war: it weakened the German Navy's attacks on allied shipping by disrupting the production of U-boats. The remarkable success in sinking allied shipping during the early months of 1943 had been halted and the tide was turning against the U-boat, with losses mounting. Production was no longer keeping pace with these losses and, after Operation Gomorrah, the number of submarines in service gradually declined, despite the introduction of new, prefabricated types that could be more quickly assembled. It has been estimated that by the end of the war total production of U-boats was 150 fewer craft than planned. The lost output from the Hamburg yards, as a result of Operation Gomorrah, was estimated at around twenty-five U-boats.

There are stories about the phosphorus bombs used in Operation Gomorrah which cannot now be confirmed. Why did Bomber Command drop so many of them on Hamburg? When I asked this question of one of the archivists at the Imperial War Museum in London, he offered the opinion that the bombs in the British stockpile were becoming unsafe and needed to be used quickly. Phosphorus bombs are difficult to store unless the bomb's tin-plate casing is absolutely air-tight. Even the slightest hole will allow air to get in and its oxygen and water vapour will react with the phosphorus to form phosphorus acids, which will eat away at the tin plate and soldered joints. The simple answer to this problem would have been to dispose of the bombs over Hamburg.

In the city itself there were stories of terrible suffering. People splattered with burning phosphorus had sought relief in the canals of Hamburg where they pleaded for help, but waited in vain, until the skin peeled away from their bodies. The only recourse was to put them out of their misery and it was rumoured that armed troops went along the canals delivering the *Gnadenschuss*, literally the 'shot of grace'. This story might have arisen from the decision to cordon off the so-called Dead Zone of the firestorm area and to allow access only to troops and police.

Martin Caidin, in his book *The Night Hamburg Died*, mentions these reports but places them at the Alster lake, outside the firestorm area. Charles Whiting in his book *The Three-Star Blitz* also recounts the story, which he heard when he entered

Germany with the advancing troops towards the end of the war and was confronted by the horrors of the devastated German cities. On the other hand Martin Middlebrook, in *The Battle of Hamburg*, mentions the story but says that several authorities in Hamburg, whom he interviewed at length, denied that it had happened. Nevertheless, it has a ring of truth to it. Many who sought to escape from the fires must have been badly burned by phosphorus and it might well have seemed humane to put them out of their misery, especially if their burns were so horrific that they were beyond hope of healing.

Nature does strange things when there has been great devastation. Pepys marvelled at the way red flowers sprang up all over the wasteland that was London after the Great Fire of 1666; the battlefields of Flanders were carpeted with miles and miles of poppies after the First World War; while, after the Second World War, the vast acres of bomb sites in London became luxuriant gardens of pink rose-bay willow herb, normally a rare wild flower of the English countryside. In Hamburg nature responded again with what was interpreted as a message of hope: in the autumn of 1943 the many lilac and chestnut trees of the city suddenly came into blossom as if it were spring.

Otherwise there was little to hope for. The war had not ended for Hamburg, although for almost a year the bombers left it alone. Then, in July 1944, the USAF came back with a series of heavy daylight raids, while on the night of 28/29 July the RAF sent 300 bombers to the city again. One of the worst raids was yet to come. On 25 October, a force of 455 American bombers set out to bomb the three oil refineries of Harburg, in the southernmost part of the Hamburg district. Because cloud obscured their targets, they switched their attack to the town of Harburg itself and destroyed most of it, killing 750 people. The final air raids that Hamburg was to suffer came in March and April of 1945, when the RAF bombed the city five times, trying to destroy the U-boat yards where new types of U-boats were being made that could stay underwater for long periods. The last raid was on the night of 13 April 1945 and the war ended three weeks later. Altogether Hamburg was hit by seventeen major raids during the Second World War, during which 16,000 tons of bombs were dropped, destroying 75 per cent of the city.

△

The allies were not the only ones to use phosphorus bombs. There were occasions when the Luftwaffe dropped them on Britain. One such bomb fell on Birmingham in an air raid during the night of Wednesday 5 August 1942. Vere Hodgson* wrote in her diary that a new kind of phosphorus bomb had fallen on the Queen's Gravy Salt factory and the place had gone up in flames. She also recorded that one fell on a block of flats called Coleherne Court, in Earls Court, London, during the middle of October 1943 when there were several small air raids on the capital.

On occasion, phosphorus bombs could be as much a threat to the bombing aircraft as to those in the target areas below, and sometimes exploded prematurely in the bomber. Those who experienced such an accident rarely lived to tell the tale but, on 12 April 1945, during a raid on Koriyama, Japan, Staff Sergeant Henry Erwin of the USAF was confronted with a faulty phosphorus bomb which ignited inside his B-29 bomber. In an act of bravery that later earned him the Medal of Honor, Erwin picked up the bomb and jettisoned it through the co-pilot's window; he suffered severe burns, but survived.

---

* Originally from Birmingham, she lived in central London throughout the Second World War and wrote long letters to her family and friends recounting her experiences. These were published in 1976 as *Few Eggs and No Oranges.*

Razzles, deckers and Fu-Gos

Among the more bizarre phosphorus weapons of the Second World War were the razzle and the decker. The razzle consisted of a layer of phosphorus sandwiched between sheets of celluloid, while the decker was larger and its sandwich also contained a filling of rubber-petrol jelly. It was said that they were modelled on an arson device used by Chicago gangsters. The razzle was designed to be scattered over fields of ripe grain, the decker over the Black Forest. They were packed in dilute alcohol and transported in sealed cans. As they were strewn over farms and forests they would dry out and the phosphorus would ignite, setting fire to the highly inflammable celluloid and generating enough heat to ignite crops and trees.

Albright & Wilson, the UK phosphorus producers, were allowed to impound all stores of celluloid in order to manufacture the new weapons, thereby ending the production of celluloid toys and toothbrushes for the duration of the war. Around thirty million razzles and half a million deckers were fabricated, even though tests in England had shown that the razzles were incapable of setting fire to cornfields and, in Scotland, that the deckers were unable to ignite forests. It was hoped that German crops and wood would be more flammable and so large numbers of these devices were indeed scattered far and wide. Apart from a few forest workers and farmers, who picked up the devices and pocketed them, setting their trousers alight as a consequence, no damage was done to the German economy and none of the secret factories that were supposedly hidden in the Black Forest were brought to light.

Such devices accounted for relatively little of the thousands of tons of phosphorus produced by Albright & Wilson's factories in England and Canada during the Second World War. Incendiary bombs took the lion's share, followed by smoke bombs, which were produced at the rate of around 200,000 a week.

In Japan a new type of weapon was devised: a balloon bomb which they called the Fu-Go. These were designed to attack the US mainland. They were launched from Japan and were carried by the prevailing winds that blow across the Pacific Ocean. They consisted of a large balloon with a weapon load of one 15 kilogram anti-personnel bomb and two phosphorus bombs which, it was hoped, would start forest fires. On 3 November 1944, the first Fu-Gos were released, and during the following six months almost 10,000 were launched, of which around 1,000 reached North America, landing in states as far apart as Alaska and Texas, not to mention Canada and Mexico.

Fu-Gos caused little damage and the only victims were some children on a church picnic. On 5 May 1945, the Rev. Archi Mitchell, of Lakeview, Oregon, and his wife Elsie were enjoying a day in the woods when they discovered a Fu-Go which, rather foolishly, they tried to drag from the woods in an attempt to prevent a fire. This triggered the 15 kilogram bomb, which exploded, killing Elsie and five children aged eleven to thirteen.

Fu-Gos were lurking in remote areas of the US and Canada for many years after the war. One was found in Alaska in 1955 and was still primed to explode if disturbed. The US Government had adopted a strategy of misinformation about Fu-Gos in an attempt to mislead the Japanese. The media agreed not to report any damage caused by them, with the exception of one incident which told of a balloon bomb reaching Wyoming and failing to explode. This limited piece of information was allowed to get back to the Japanese in the hope of convincing them that their secret weapon had been a failure.

After the Second World War was over, the British Government wondered what to do with its stock of unused phosphorus bombs. In the end it decided to dump them in the sea, in a deep channel lying several miles off the Scottish coast and to forget about them, hoping that slow corrosion would dispose of them. It came as a shock therefore to find them washing ashore fifty years later. In 1995 locals walking on the beaches near Beauforts Dyke reported finding curious 'devices' which were rusty and decayed but nevertheless looked dangerous. Defence experts were

---

Modern phosphorus weapons

The US armed forces use phosphorus-filled smoke cartridges as spotting and marking rounds, thereby enabling a better aim for mortar shells. The M722 60 mm smoke cartridge, weighing 1.68 kg (3.7 lb), consists of a detonating fuse and a charge of explosive which shatters the casing and disperses the phosphorus, which in turn immediately produces clouds of smoke. Depending on the propelling charge, the M722 has a range of up to 3 kilometres and can be fired from a mortar at the rate of twenty rounds per minute.

The M929 120 mm smoke cartridge is a much more powerful weapon, weighing 14 kg (30 lb) and with a range of 7 kilometres. The first US Army unit was equipped with these shells in 1995. They are loaded with felt wedges impregnated with phosphorus and are dispersed by the bursting charge. Some M929 shells are produced with delay fuses.

The US Army also has a much larger 155 mm shell, which is fired from an M198 Howitzer. It is mainly an incendiary device which delivers 14 kg of phosphorus. In addition there is the M34 hand grenade which contains phosphorus, as does the M15, a rifle-launched grenade. Tests on the hand grenade, which were undertaken in 1955, showed that it has a dispersal pattern of 20 metres in radius (about 70 feet) and that it produces a cloud of thick white smoke as well as showering the area with burning phosphorus particles.

called in to inspect them and identified them as 30 lb incendiary bombs. Teams of divers from the Clyde Submarine Base were then sent to search for other bombs that might have been shifted by undersea currents from their supposedly final resting place.

△

In the history of phosphorus its role in the bombing of Hamburg is among the most shocking and it is almost inconceivable today that such episodes would ever be repeated. In the two world wars incendiary bombs were made from all kinds of materials, but nothing compared to the fires produced by exploding and burning phosphorus, against which there was no defence for buildings or people. Dropping phosphorus bombs on the homes of innocent civilians is never likely to happen again, yet phosphorus weapons are still being manufactured, as the box shows. As long as wars are fought they will be needed because no other substance can produce the dense smoke of phosphorus pentoxide that burning phosphorus quickly creates. For this reason phosphorus will continue to be part of the armoury of all modern armed forces into the foreseeable future.

The Alchymist in Search of the Philosopher's Stone, Discovers Phosphorus. Painted by Joseph Wright A.R.A. of Derby (1734–97). *(The Bridgeman Art Library)*

Robert Boyle (1627–91). Ambrose Godfrey (1660–1741).

Apparatus for the preparation of phosphorus,
Ambrose Godfrey's phosphorus factory c.1720.

Above: An advertisement for Darkest England matches, 1894. *(Salvation Army)*

Left: Johann Lincke, the German apothecary who first sold phosphorus as a medical treatment (1675–1735).

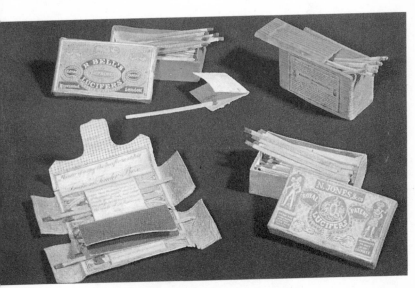

Some early types of lucifer match.

Left: Annie Besant (1847–1933). *(Salvation Army)*

Below: The matchgirls at work in the Bryant & May factory. *(Salvation Army)*

Right: Dipping the matches at Bryant & May, a primary cause of 'phossy jaw'. *(Salvation Army)*

Below right: A jawbone that clearly shows the effects of 'phossy jaw'. *(Salvation Army)*

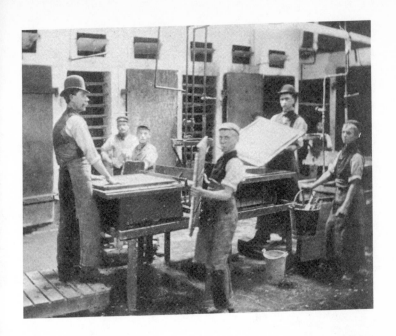

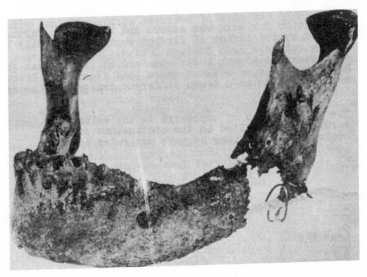

Left: A phosphorus fire at an Albright & Wilson warehouse in Bristol, 1990. *(Albright & Wilson)*

Below left: A phosphorus shell exploding. *(Albright & Wilson)*

Right: A phosphorus bomb dropped onto a ship. *(Albright & Wilson)*

Below: Louisa and Alfred Merrifield, who poisoned their employer with doses of phosphorus. They were tried for murder in 1953.

Right: Mary Wilson, a skilful poisoner, who murdered at least two husbands using a phosphorus-based beetle poison.

Below: The Placentia Bay phosphorus plant, Newfoundland. *(Albright & Wilson)*

8. The ultimate evil – and a power for good

It may be several weeks, or even months, before I shall ask you to drench Germany with poison gas and if we do, let us do it one hundred percent.

Winston Churchill pronounced these words to his Chiefs of Staff in 1944. Fortunately for the allied forces, they were never acted upon. Had they been, retribution would have been far worse than Churchill could ever have imagined, because the Nazis had stores of thousands of tons of a secret chemical warfare agent, far more toxic than anything previously known and against which there was no defence. That agent was the nerve gas tabun, a new type of phosphorus compound many times more deadly than the mustard gas Churchill could deploy.

What prevented Hitler from using his new chemical weapon? By 1944, his position had become desperate, his forces were retreating on all fronts and his cities were being pounded to rubble. His generals and military advisers urged him to use tabun, but Hitler stayed his hand and refused to play his one last ace, believing that the allies must also have stocks of nerve gas and would retaliate in kind. Ironically, he refrained from using the one weapon that might have saved his crumbling empire.

Could tabun really have influenced the outcome of the war in Europe? Perhaps not, but it would have caused panic wherever it was used, whether on the battlefield or in cities. One breath of tabun causes temporary blindness, two lungfuls incapacitate, while three will kill within a few hours.

Phosphorus, in the hands of Nazi chemists, had finally plumbed the depths of evil. Yet chemicals that resemble nerve gases, the organophosphates,* are today in use around the world to increase food production; and they are among the most effective pesticides we have. Despite the undoubted benefits they bring, some groups campaign against them, believing that they cause chronic ill health, and cite their relationship with nerve gas as a good reason to be suspicious of them. This suspicion is understandable, though it clouds the debate about the safety of pesticides. The link with nerve gases can be traced back to the time when chemists started researching into organophosphorus compounds.

The N-Stoff of the Nazis

In the First World War more than 110,000 tons of chemical warfare agents were deployed against troops, sometimes with spectacular results. The first gas attack came on the morning of 22 April 1915, when the Germans released a cloud of chlorine gas towards the British forces on the Western Front. Weather conditions and the prevailing wind were ideal for carrying the choking gas across no-man's land and into the British trenches.

---

* Strictly speaking we should differentiate between an organo*phosphorus* compound, in which there is a carbon-to-phosphorus bond, and an organo*phosphate* compound, in which the phosphorus atom is at the centre of four surrounding oxygens to which the carbons of the organic groups are attached. The convention now is to call them all OPs, but this term has become rather judgmental, and by implication all OPs are dangerous. As we shall see, this is not necessarily so. Indeed all living things depend on OPs; technically speaking DNA is an OP.

There, 5,000 men died in agony and 15,000 were wounded. Later in the war, other toxic gases were used, such as phosgene, mustard gas and Lewisite,* but these were countered by equipping troops with better gas masks and protective clothing and providing effective treatment for those who had been exposed to them.

At the time the use of chemical weapons had been completely unexpected because they were outlawed in the Hague Declarations of 1899 and 1907, which drew up guidelines regarding conduct in war. Thereafter and despite paying lip-service to the aims of these declarations, all the major powers made war gases and used them. Research into these deadly chemicals continued after the Armistice in November 1918 and, with the growth of air power, it was envisaged that in future wars even cities would be attacked with gas. This threat was taken so seriously that, when the Second World War seemed inevitable in 1938, all civilians in Britain began to be issued with gas masks, including infants, who were given Mickey Mouse versions made from coloured rubber. Thankfully, they were never needed, but the danger was always there.

The enormity of that danger only came to light after the war, when the armies invading Germany came across massive stockpiles of tabun and other nerve gases. As the Russians advanced from the east in 1945 they stumbled upon a heavily protected chemical plant at Dühernfurt near Breslau in Prussian Silesia (today Dühernfurt is known as Dyhernfurth and Breslau as Wroclaw and both are in Poland). It was eventually discovered that the plant had been manufacturing a new chemical warfare agent, tabun, at the rate of 1,000 tons per month. The plant had worked flat out during 1944, until allied bombing of the German

---

* Although they are called gases, most chemical warfare agents are in fact liquids and the word gas actually refers to the vapours given off by these liquids.

chemical industry had cut off supplies of methanol and cyanide, which were needed to produce tabun, so production had virtually ceased by the time the Soviet forces arrived. The Russians subsequently dismantled the plant and shipped it back east, where it was again put into production.

Meanwhile, the British and American forces were advancing from the west and they came across huge stockpiles of 105 mm shells charged with an unknown agent. There were tens of thousand of these shells containing a total of around 12,000 *tons* of a chemical they did not recognize. Again it was tabun. More than 70,000 such shells were shipped back to the UK – and they were eventually disposed of in the deep Atlantic ten years later.

To begin with, the British thought the new weapon might be a blistering agent like mustard gas, which, when it comes in contact with skin, causes terrible wounds that refuse to heal. So they put several drops of the new chemical on to the forearms of volunteers but surprisingly no blisters appeared as the drops evaporated. Nevertheless, it was clear that something was happening because investigators and volunteers suddenly found they could hardly see, their noses began to run and their heads began to throb.

The semi-blindness caused by the new liquid puzzled the investigators, so they put a drop in the eye of a rabbit to see what effect it had. The animal promptly died. Next, they took groups of subjects and asked them to take a deep breath of air containing a little of the vapour, enough to supply only a fraction of a milligram of tabun. The effects were again dramatic: blindness, vomiting, breathlessness and convulsions. These preliminary tests were halted before anyone died. Without doubt, the Germans had found a chemical agent that was far more powerful in disabling people than any other war gas.

Unlikely as it seems, the Nazi chemists who originally discovered the nerve gases were well intentioned. They were looking to make better pesticides and worked for IG Farben, the giant

German chemical combine and the world's largest chemical company. The team was headed by Gerhard Schrader and, in the early 1930s, they made and tested a variety of sulphur compounds with little success. Although some of the chemicals they made were deadly to insects, they were not suitable for commercial exploitation. Then, in 1935, Schrader turned his attention to phosphorous–fluorine compounds and quickly discovered that these were highly toxic, *too* toxic in fact for commercial use. He was not the first to make these kinds of molecules – they had been reported in technical journals as long ago as 1902 – but he was the first to realize that their toxic power might be useful for something, and a patent for them was filed by him on behalf of IG Farben in 1939.

One of the new compounds was more toxic than anything Schrader had ever come across before: its chemical name was 1-methyl-2,2-dimethylpropyl methyl phosphonofluoridate. Schrader decided to call it sarin, a name devised from letters in the surnames of the chemists in his research group: *S*chrader, *A*mbos, *R*üdringer and Van der L*in*de. Animal tests revealed how deadly sarin really was. The company passed the information on to those engaged in chemical-warfare research, who immediately classified the discovery as top secret and gave the new chemical their own code name 'N-Stoff'. They tested it not only on insects but on guineapigs and even on apes. It had all the hallmarks of a superb chemical-warfare agent, and a pilot plant was constructed at Münster-Lager for its manufacture.

Soon, other equally deadly phosphorus compounds came to light and these were named soman and tabun. Tabun is a phosphorus–cyanide compound which is not only deadlier than sarin but much easier to make, and it went into large-scale production in 1942. By the end of the Second World War the Nazis had manufactured enough tabun to exterminate all human life on Earth. (Only a milligram is needed to kill someone, and the Nazis had in their possession enough tabun to exterminate twelve *trillion* people.) It had even been tested on concentration-camp inmates, but those data were destroyed by the staff at IG Farben's headquarters in Frankfurt in the closing weeks of the war. Details only leaked out during the Nuremberg Trials.

Among those tried as war criminals was Albert Speer, who was Minister for Armaments and War Production: he was found guilty of using slave labour. Unlike most of the Nazi leaders, he was not executed but given a long prison sentence. In his memoirs, *Inside the Third Reich*, he says he seriously considered killing Hitler in 1945 by releasing nerve gas into the ventilation shaft of the Führer's underground bunker in Berlin. His plan was thwarted when the inlet shafts were suddenly raised several metres to guard against possible gas attacks.

It is less well known that the British were also researching organophosphorus agents as potential war gases. A team of Cambridge chemists, led by Dr Bernard Saunders, was directed to work on their development in the early 1940s, maybe as a result of a tip-off from within Germany. Even without such a pointer, Saunders knew from reading the chemical literature that one compound, diethyl phosphorofluoridate, had been made in 1932 by the German chemist Lange, who claimed it was potentially very dangerous. They were not interested in this aspect of the chemical and took the research no further.

One member of the team working for Saunders said that the Cambridge team decided to investigate this and similar compounds in the belief that, because both phosphorus and fluorine were toxic, a molecule that brought them together should be even more toxic – and they were right but for the wrong reasons.*

Saunders recruited a research team of brilliant young chemists and began to make compounds similar to diethyl phos-

---

* In fact phosphorus and fluorine can cancel out each other's toxicity, as they do in sodium fluorophosphate, which is added to toothpaste.

phorofluoridate. Some of the compounds they made were toxic, but many were not. Some appeared to be excellent at killing insects but had little effect on mammals. One was especially toxic, diisopropyl phosphorofluoridate (DFP), which was deadly to rats at a concentration of only 0.36 parts per million in air, when breathed for ten minutes.

Saunders even tested some of it on himself and a volunteer. They sat in a sealed room, put a drop of the vapour on a glass dish and then simply waited. After an hour, during which nothing appeared to have happened, an assistant went into the room to ask Saunders how he was feeling; Saunders replied that he felt very strange and had been wondering why the room had gone dark! (Saunders was unaffected by this experience and lived to be eighty.) He describes such tests and the work that was done at Cambridge on nerve gases in his 1957 book *Some Aspects of the Chemistry and Toxic Action of Organic Compounds Containing Phosphorus and Fluorine*, a title that belies its dramatic contents, covering as it does all aspects of nerve gases, their effects and antidotes.

Dangerous though it was, however, DFP was not as toxic as some war gases then known, so it did not justify diverting resources into production on a large scale, and the British continued to make their preferred chemical agent, mustard gas. Even so, patents were taken out on the compounds that Saunders had made in 1943 and 1944 and these were eventually published in 1948 as British patents nos 601,201 and 602,446.

After the war, the British transferred equipment from the Münster-Lager sarin plant to Nancekuke in Cornwall and carried out research there. They also manufactured the Nazi nerve gases and produced around 20 tons but all were destroyed when the Government took a unilateral decision to abandon chemical and biological weapons in 1956. The sarin they had made was rendered harmless by treating it with sodium hydroxide solution (caustic soda), which breaks the phosphorus–fluorine bond.

How nerve gases work

As their name implies, the nerve gases disrupt the body's nervous system and they do this at locations where nerve impulses travel down nerve fibres and then trigger the release of special messenger chemicals which carry the message forward to its destination.* The chemical messenger that transmits the signal across the neural junction is acetylcholine and, after it has activated the muscle, gland or another nerve, it has to be deactivated by an enzyme called acetylcholinesterase (AChE for short). This splits the acetylcholine into its constituent parts, acetyl and choline, which then find their way back to the starting point, where an enzyme waits to recombine them, in preparation for carrying the next message.

If there was no AChE to neutralize the acetylcholine, it would endlessly keep activating the muscle, gland or nerve. So vital is this enzyme that were anything to interfere with it at key organs, such as the lungs and the heart muscle, then death would quickly result. Nerve gases do just that.

The immediate symptoms of nerve-gas poisoning are frightening because they undermine the body in three ways. First, they attack the central nervous system resulting in giddiness, anxiety, headache, confusion, convulsions and difficulty in breathing. Second, they disrupt the regulation of the various glands of the body and this leads to a runny nose, copious phlegm, profuse sweating and excess salivation. Third, they cause muscles to respond erratically, which results in miosis (closing of the iris of the eye), stomach cramps, diarrhoea, incontinence and irregular heartbeat.

---

* This is analogous to the nineteenth-century method of sending messages, first by wire to a local post office, where it would be printed out and then taken by hand as a telegram to the recipient.

Nerve agents are able to deactivate AChE by reacting with it chemically, in other words binding themselves to the enzyme. How the nerve junctions are affected varies according to the nerve agent, but all of them act first on the muscle that controls the iris of the eye. Thus the iris closes and everything suddenly appears to have gone dark. As more of the agent circulates within the body, other parts of the nervous system are affected. Meanwhile, the body is trying its best to remove the offending molecules as fast as it can, but the battle is quickly lost and, unless outside help is at hand in the form of certain drugs, death is the likely outcome. The symptoms suffered after poisoning by nerve agents are really those of acetylcholine accumulation, and the way a nerve gas affects people depends on the rate at which it deactivates AChE and the rates at which the poisoned enzyme can be rescued.

The body's main lines of defence are other enzymes known as paraoxonases, whose job it is to free the blocked AChE, but they are limited in what they can achieve and there are not enough of them to cope with a massive disruption of AChE, mainly because they have not evolved to deal with such a situation. Only now are we beginning to appreciate that the paraoxonases can vary as well, as revealed by Bert La Du of the University of Michigan Medical School in a paper published in *Nature Medicine* in 1996. It may well be that a person's paraoxonases will determine how they respond to nerve gases and even to organophosphates, which could explain why some people are more sensitive to them.

This appears to have been the case with one human guinea pig, a twenty-year-old soldier doing his National Service, Lance-Corporal Ronald Maddison who was an RAF mechanic stationed at Ballykelly in Northern Ireland. The full circumstances surrounding his death were made available in the summer of 1999 when documents in the Public Records Office were declassified.

At 10.15 a.m. on 6 May 1953, Maddison and five other men

were seated in a sealed chamber in the Chemical Defence Experimental Establishment at Porton Down, Wiltshire. A piece of material from which uniforms were made had been taped to their arms and on to this was dripped 200 mg of sarin, the idea being to see to what extent this would penetrate clothing and enter the body through the skin.

After about twenty minutes Maddison complained of feeling queer, began sweating profusely and within minutes he had passed out. He was quickly removed from the test chamber and given medical treatment but by now his face was ashen and he had lost control of his bowels. By 11 a.m. his heart had stopped beating. He was rushed to Porton Down hospital and attempts to resuscitate him were made for over two hours, but to no avail.

According to other National Servicemen at the time, Maddison volunteered to take part in the tests in the belief that these were to do with finding a cure for the common cold. (At the time, the Government's Common Cold Unit was also in Wiltshire and it too advertised for volunteers. It has since closed down.) Soldiers were encouraged to come forward on the understanding that they would receive extra pay and leave.

Although hundreds of servicemen took part in nerve gas tests, only Maddison (test subject no. 702) died but, as a result, a government inquiry into his death ruled that in future no more than 5 mg of sarin could be used at any one time in such trials. However, the amount that Maddison was exposed to was much less than the 1,800 mg that is considered the lethal dose via clothing and the skin. It seems likely that he was just unlucky, in that his body's metabolism was ultra sensitive to AChE inhibition and as a result his heart muscle quickly ceased to function.

The ultimate nerve gas, VX

Despite reports to the contrary (see box) no democratic power has ever used nerve gases in war. In the years that followed the Second World War, a great deal of work went into making and testing organophosphorus nerve gases in government laboratories around the world and into making and testing organophosphate insecticides in industrial laboratories. It was at the British chemical company, ICI, that Dr Ranajit Ghosh discovered an even deadlier kind of organophosphorus compound, ethyl S-2-diisopropylaminoethyl methylphosphrothiolate, which was eventually given the slightly easier-to-remember code-name VX. This was the first of the V-agents and it was immediately classified as top secret and taken for testing as a chemical weapon. Its

---

Did the US use nerve gas in the Vietnam War?

In its issue of 15 June 1998, *Time* magazine stated that US armed forces had used sarin in September 1970 against a village in Laos, where GIs who had gone over to the Viet Cong side in the Vietnam War were believed to be hiding. Nor was this the only occasion in that terrible war when the US military had used sarin, the article said. They had also used it against the Viet Cong when rescuing pilots who had been shot down.

None of this made sense. No military commander would use nerve gas in such cavalier fashion and still hope to achieve the stated objective. In fact, the story was nonsense and, within days, *Time* magazine had recanted and apologized. The story had been written by a producer from the TV news channel CNN – which is also owned by Time–Warner, the publisher of the magazine – and there heads rolled.

manufacture was secretly patented on 13 August 1962, and these patent documents were declassified in 1974, leading to vociferous protest, and they were immediately removed from all UK patent libraries. Unfortunately this did not prevent those copies that had already been sent to patent libraries around the world from reaching their destinations. There they were beyond the control of the British Secret Services and, in one place at least, they fell into the wrong hands.

Four V-agents were considered as potential nerve gases: they were coded VX, VE, VG and VM, and all have a phosphorus–sulphur bond, with the sulphur as part of a complex group that contains nitrogen. When animal tests revealed how deadly these were, it seemed impossible that they could ever be tested on humans, but a volunteer came forward, Dr Van M. Smith. In the first test he was injected with around 2.5 micrograms of VX and appeared to suffer no ill effects. Three hours later, he was given double this amount and his body began to react. He had a headache, started sweating, felt light-headed and suffered stomach pains, yet his blood cholinesterase was normal.

A year later, he was allowed to take part in a further test which this time involved being injected with around 15 micrograms of VX. His cholinesterase fell to two-thirds of the normal level, but he still appeared to be OK. After two hours, more VX was given at the rate of one microgram per minute and then serious effects began to appear: he became pale, stopped talking, had profuse salivation and started vomiting. By now, his cholinesterase was down to 15 per cent of the normal level and the experiment was stopped.

Other tests were carried out on the ability of VX to penetrate the skin, by using radioactive phosphorus to make the VX, and it was found that it could enter the body this way. Others drank water containing VX, and water that had been passed through a field kit designed to filter out nerve gases. The onset of symptoms was immediate for those who drank the untreated water, but what surprised the scientists monitoring the tests was that individuals who drank the filtered water also showed similar symptoms, although these did not appear for several days. In 1969 there was a leak of VX from a plant at Newport, Indiana, where US nerve agents were being manufactured. The vapour drifted across the aptly named Skull Valley, where a flock of sheep was grazing, and killed all 6,000 of them.

While considerable research was devoted to the manufacture of nerve gases, even more went into ways in which they might be countered and how those affected by the gases might best be treated. The results of this research had wide implications because treatments were developed that would help those affected by organophosphate insecticides.

As explained earlier, nerve gases produce a variety of symptoms. How would soldiers behave if they were suffering from some of these? A group of 134 enlisted men and officers of the US army took part in an exercise in which they were exposed to sub-lethal battlefield levels of sarin. Almost all found their sight greatly diminished, three-quarters suffered runny noses and half had terrible headaches and difficulty in breathing. They were asked to carry out certain manoeuvres, including firing their rifles. The men became confused and depressed and their accuracy and speed of fire was below par, while their officers became easily irritated and flustered. However, it was concluded that they remained an effective fighting force – provided they were not being asked to fight at night. After dark, they were unable to see anything.

Experiments showed that it was even possible to remove a gas mask to eat and drink while exposed to an atmosphere of nerve gas – provided you were quick about it. Those who took only a few seconds to have a drink generally came to no harm and even those who took over a minute to eat some food only experienced mild symptoms. The skill was to take several deep breaths before removing one's gas mask.

Table 8.1: Deadly doses of nerve agents (approximate values)

Nerve agent	Toxic dose via skin (mg/kg)*	Toxic dose when inhaled (mcg/kg)**	Toxic dose when injected (mcg/kg)**
Sarin	25	90	2
Soman	20	70	n.a.
Tabun	25	n.a.	15
VX	0.1	5	less than 2

* Milligrams per kilogram body weight.

** Micrograms per kilogram body weight. (A microgram is a millionth of a gram and weighs about as much as a fleck of dust.) The figures in the table should be multiplied by seventy to estimate the amount that would be fatal for an average person. Thus the total amount of VX that is fatal when breathed in is around 350 mcg, which is 0.00035 gram.

It is not easy to assess the amount of nerve gas that will prove fatal to an average person. Tests on those exposed to tiny amounts give some idea of the doses that can kill, and there were several incidents at plants making nerve gases in which individuals were suddenly exposed to large amounts, although how much was not easy to judge. Within seconds, people would lose consciousness and go into convulsions, but prompt action invariably saved their lives and all were able to resume work after a few weeks. Data gathered from these incidents make it possible to assess what would constitute a lethal dose of the four key nerve agents, and these figures are given in Table 8.1.

Sarin attack!

Sarin is a colourless liquid which boils at 147°C. Its vapour pressure at 20°C is 2 mmHg, enough to deliver an incapacitating dose when breathed in for only a few minutes. It can be made from phosphorus trichloride (PCl_3) which is reacted first with chloromethane (CH_3Cl) to attach a methyl group (CH_3) to the phosphorus in place of one of the chlorines, then with hydrogen fluoride (HF), which replaces a second chlorine with fluorine, and then with propan-2-ol (C_3H_7OH), which replaces the third chlorine. The chemical reactions which take place are more complicated than this simple sequence suggests and the phosphorus also acquires an oxygen atom, but the result is sarin: $(C_3H_7O)P(O)(CH_3)F$.

Sarin is easier to make from the commercially available chemical methylphosphonic dichloride, but sales of this are carefully monitored around the world for this reason. Because of such restrictions it is difficult to obtain supplies of the ingredients necessary to make sarin, but a skilled chemist could, given time, make them from quite innocuous materials and produce sarin in kilogram quantities, in other words enough to kill or disable tens of thousands of people. Indeed, a tablespoonful of sarin would be enough to wipe out a town of 25,000 inhabitants if it were sprayed from the air as a fine mist.

Table 8.1 shows that sarin is the least toxic nerve gas, which may be why it has been used in recent years since it exposes its makers and users to less risk. There have been three recorded incidents in which sarin has deliberately been dispersed among innocent people: once in Iraq and twice in Japan.

In 1988, Kurdish villagers claimed they were victims of a gas attack by the Iraqi Government which left many women and children dead. Tabun was the suspected agent since it is easy to synthesize. While it was strongly suspected that they had been exposed to a chemical agent, it was impossible to prove at the time which agent had been used. It has since been discovered that the Kurds' accusations were true – but the agent was not tabun, it was sarin. In 1993 James Briscoe, an archaeologist working for the US group Physicians for Human Rights, collected soil and shrapnel from bomb craters in Kurdish villages. These samples were taken to the British Chemical and Biological

Defence Establishment at Porton Down, and chemists found traces of substances derived from sarin in the soil. They even detected sarin itself on a bomb fragment, where its absorption into the layer of paint had preserved it.

Twelve people died and more than 5,000 were injured when members of the Aum Shinrikyo doomsday sect released sarin on the Tokyo underground during the morning rush hour on 19 April 1995. An earlier release of sarin by the same group had killed seven people and injured 200 in Matsumoto, Japan, the previous year. That attack alerted the authorities to the terrorist threat and so, when trains on the Tokyo underground were attacked that fateful morning, the symptoms of the victims were soon identified as sarin-related. Many lives were saved because the antidotes were on hand and thousands of those affected were able to return home after a few hours. The most common symptoms were miosis, headache and breathing difficulties. For most of the victims, the miosis was long-lasting and very painful. Shoko Asahara, the leader of Aum Shinrikyo, the religious cult whose followers carried out the attack, is now in a Tokyo prison, but the cult continues to attract members and it is fabulously wealthy, thanks to its investment in the mini-computer business.

VX was also produced in the laboratories of Aum Shinrikyo, which in the early 1990s were located near Mount Fuji. Cult agents used VX to dispose of one former member, a twenty-eight-year-old man who had fallen foul of the organization. He was stopped in the street in Osaka in December 1994 and, while one man confronted him, his associate squirted the victim with VX on the back of his neck. As they ran away, the man gave chase but collapsed after a few steps and went into a deep coma from which he never recovered. He died ten days later.

At the time the doctors at the hospital where he was treated thought he had been poisoned with an organophosphate insecticide, but the real cause of death came to light when members of the cult, who were arrested after the Tokyo subway attack, admitted using VX to kill him. When samples of the dead man's blood were sent for further analysis, chemists confirmed the presence of two compounds that could only have come from VX.

Antidotes for nerve gases (and other OPs)

A combination of antidotes is needed to deal with the effects of a nerve gas – and to treat those suffering from any form of organophosphate poisoning. The antidotes are atropine, oxime and diazepam (this last drug is better known as an antidepressant under the trade name Valium). The Germans discovered the antidote effects of atropine during the Second World War. Oxime antidotes were identified in the 1950s, and a combination of these with atropine was found to be more effective against sarin and VX when used together than when used alone.

Atropine itself is a deadly poison, which normally kills by blocking acetylcholine receptor sites. However, when these receptors are being mercilessly bombarded by acetylcholine, as they are when AChE has been deactivated, it is necessary to switch the receptors off, and this is what atropine can do. The atropine provides immediate though temporary relief. But it doesn't solve the real problem of unblocking the enzyme so that it can resume its normal function.

This unblocking is carried out by oximes, such as N-methyl-pyridinium-2-aldoxime (pralidoxime), which can be administered by injection or taken by mouth. Oximes have the chemical group NOH, which reacts with the nerve gas and thereby destroys its ability to interfere with AChE. Once free, the AChE can again to do its work of removing acetylcholine.

The third drug, diazepam, counters the over-stimulation caused by nerve gases, and it is given to reduce apprehension, agitation, muscle spasms and convulsions.

The modern soldier can be equipped with an antidote package that provides excellent protection against nerve-gas attacks. When such an attack is expected, he can inject himself with the antidote using a special field kit to penetrate his battledress and protective gear.*

However, nerve gases are being boycotted worldwide and soon there may be no need for such protective and preventative measures. The US Army's chemical weapons disposal facility at Tooele, Utah, has a special incinerator costing $600 million for the purpose of destroying these weapons. More than 10,000 tons of chemical warfare agents are stored at the facility awaiting disposal.

Organophosphate insecticides (OPs)

Sarin itself was once tested on the plant-louse phylloxera, which attacks grapevines. A 0.1 per cent solution of sarin was spread on the ground near the vine root and completely eliminated the infestation. Effective though it was, sarin was far too dangerous to be used as an insecticide.

During the course of Schrader's research into nerve gases his team came across several other OPs that were eventually to find use as insecticides, such as tetraethyl diphosphate and parathion. After the war, tetraethyl diphosphate was manufactured by the US chemical company Monsanto, but it was rather too toxic for general use and, after a series of accidents, was withdrawn. The same happened with parathion, but a variant of this, methyl

---

* In war, nerve gases would be used primarily to prevent an enemy occupying key sites, such as airstrips or port facilities. Since nerve gases have no detectable smell it would be possible to absorb a lethal dose before being aware of them. The British Army advised its soldiers to listen for the birds; if they could be heard, there was no nerve gas around.

parathion, was found to be much safer, although it is banned for indoor use. Its toxicity for humans has been estimated to be around 50 mg/kg, which means that for the average 70 kg adult as little as 3.5 g could be fatal.

Outside it is safe because it breaks down under the effect of the ultraviolet rays of the sun, but indoors it can persist for years. This is rather unfortunate because it is probably the best insecticide to use on cockroaches, a common household pest in the USA. In the southern states, some rogue roach-control agents started using methyl parathion because it was so effective, and they found a willing market for their services. By 1997 they had sprayed almost 3,000 homes in Mississippi and neighbouring states. All needed to be decontaminated and, to prevent further abuse of methyl parathion for such purposes, the Environmental Protection Agency (EPA) persuaded the manufacturer to add an odourizing agent, the foul-smelling valeric acid, to the insecticide.

When IG Farben's research was declassified at the end of the Second World War, and it was realized what advances had been made in finding potential pesticides, chemical companies around the world began to look at other OPs for possible insecticide use. Several companies already had a selection of phosphorus compounds available for testing, compounds that had been made with other uses in mind, such as oil additives. All kinds of compounds were tried on insects and some, proving to be highly effective at killing them, were marketed. But people were unaware of the chemical minefield they were walking into when they started working with OPs. Several research workers died as a result of their exposure to these compounds, as did those who sprayed them on farms and in fruit orchards.

One chemical which came to light during this search was eventually to be a big money-earner: malathion. This had originally been made as a flotation agent – used in mining to separate ores from crude rock – but it had not proved suitable. During tests it was discovered to be deadly to all kinds of insects but harmless to mammals. Unfortunately it smells of bad eggs due to the sulphur atoms it contains, which is why it is generally only used by gardeners. But it has occasional indoor use in shampoos formulated to kill head lice.

The success of malathion encouraged others to make similar, safer OPs and today these are employed as pesticides throughout the developed world.* Glyphosate is one of the most widely used herbicides. Table 8.2 lists some of the more common OPs in the order in which they were introduced. Heading the list are the two that came to light when the US Field Information Agency and the British Intelligence Objectives subcommittee released IG Farben's research files after the Second World War.

OPs are cheap and easy to make and at the time appeared to offer an environmentally safe alternative to the organochlorines, such as DDT, which are effective but rather persistent. The OPs could be tailored so that they would quickly degrade in the open air, and this was achieved by incorporating a group into the molecule that would easily break away on exposure to strong sunlight, rendering the residue completely harmless. But while the environment and food crops benefited, the people who used the pesticides were at risk. It was not long before some users were suffering symptoms related to reduced AChE levels such as headaches, nausea, vomiting and diarrhoea.

Clearly insecticides such as parathion were too potent: we can infer this from the information in Table 8.2 in the column headed 'LD_{50} (rats, oral)'. This stands for 'the lethal dose that will kill 50 per cent of rats who were given it orally' and it is measured in terms of the weight of toxin in mg, per body weight

---

* Those used in the USA are mainly methyl parathion (3,000 tons per year in 1996), phorate (2,200 tons), malathion (1,700 tons) and azinphos-methyl (1,200 tons). In 1999 the US Environmental Protection Agency placed restrictions on the use of methyl parathion and azinphos-methyl being applied to crops such as apples, peaches, etc.

Table 8.2: OP pesticides

OP	Patentee	Year patented	Pesticide uses	LD_{50} (rats, oral)*
Tetraethyl diphosphate	IG Farben	early 1940s	general	1
Parathion	IG Farben	early 1940s	mites	8
Methyl parathion	not patented	1950	general	20
Dimethoate	Cyanamid	1950	systemic	250
Malathion	Cyanamid	1951	head lice, mites	1,000
Azinphos-methyl	Bayer	1956	broad-spectrum insecticide	11
Diazinon	Geigy	1956	broad-spectrum insecticide	270
Phorate	Bayer	1956	broad-spectrum insecticide	2
Phosmet	Stauffer	1956	broad-spectrum insecticide	135
Dichlorvos	Shell	1960	houseflies, fleas	65
Fenthion	Bayer	1961	mites	230
Pirimiphos-ethyl	ICI	1970	soil pests	140
Glyphosate	Monsanto	1972	herbicide	4,870
Propetamphos	Sandoz	1973	parasites	82
Phosphinothricin	Hoechst	1977	herbicide	1,800

* Milligrams of toxin per kilogram body weight (mg/kg).

in kg. Rats make good subjects for testing because they are unable to vomit, so anything they ingest must pass through their bodies. The LD_{50} test is less used these days, but at one time it was mandatory to carry out such tests before any chemical could be released for general use.

Somewhat surprisingly, not all OP insecticides are man-made.

Nature got there first. At the foot of Table 8.2 is phosphinothricin, which is superb against weeds and is produced by certain microbes. Its chemical synthesis was patented by Hoechst in 1977 and toxicity tests showed it to be relatively safe, with an LD_{50} of 1,800 mg/kg. Phosphinothricin is an amino acid with the chemical formula: $CH_3P(O)(OH)CH_2$–$CH_2CH(NH_2)CO_2H$, in which dimethyl phosphinic acid is joined (–) to the amino acid alanine. What is surprising about this molecule is that there is a direct phosphorus-to-carbon bond (CH_3–P), something which it was thought could not occur in nature but something that is intrinsic to nerve-agent OPs. Phosphinothricin is a non-selective herbicide that acts by blocking the synthesis of particular amino acids that are crucial to the life of the plant. Another naturally occurring OP is anatoxin, which is produced during algal blooms and which can devastate other forms of marine life. This too wreaks havoc as an inhibitor of AChE.

The most popular OP for controlling weeds, especially difficult ones such as couch grass, is glyphosate, which has the chemical formula $(HO)_2P(O)CH_2NHCH_2CO_2H$. This is much safer than most OPs, as its LD_{50} of 4,870 mg/kg in Table 8.2 shows, which means it would probably require around 340 g (12 oz) to kill a human. This is the herbicide of choice not only for gardeners but for those who are trying to restore wildlife habitats around the world because it is not poisonous to fish, birds, bees or earthworms.

The difference between insecticide OPs and nerve gases is the tenacity with which they bind to AChE. Nerve gases do this strongly and are difficult to remove, whereas OP insecticides are only weakly attracted to the enzyme and are relatively easily dislodged. Of course if the body is swamped with such OPs, however mild their action, it will be unable to cope, and this is what happens to those who use such insecticides to kill themselves. For those who want to kill only pests or weeds, they pose no threat if the guidelines for their use are correctly followed.

According to Graham Matthews, Professor of Pest Management at Imperial College, London, and author of the authoritative *Pesticide Application Methods*, published in 1992, this does not mean that OP insecticides are safe for everybody to use. He says that some people appear to be particularly sensitive to OPs and this is to be expected for any condition that relies on enzyme behaviour within the body. Some people have higher levels of enzymes and so are better able to detoxify unwanted chemicals in the body. Those with the lowest levels will clearly be most at risk. How many people come into this category is difficult to assess but it may be as high as one person in 200 if the numbers affected by sheep dipper's flu are anything to go by.

Gulf War Syndrome, sheep dipper's flu – and chapattis

The World Health Organization says that OP pesticides are responsible for one million serious accidental poisonings a year and that as many as two million people attempt suicide with them. Many more may be affected accidentally but only experience mild symptoms. Two groups of people have suffered illnesses which they have blamed on OP insecticides: Gulf War veterans and sheep farmers.

The 700,000 troops sent to fight the Gulf War in 1991 were thought to be at risk, not only from chemical warfare agents but from malaria-carrying mosquitoes. To counter the latter threat, all living quarters and tents were sprayed with OPs. Some soldiers claim that everything was 'drenched' with insecticides. Malathion was used to protect Iraqi prisoners of war, while allied forces had their sleeping quarters sprayed with the less foul-smelling but equally effective methyl parathion and diazinon.

Tens of thousands of US troops developed Gulf War Syndrome in the years that followed, as did 750 British soldiers, with such symptoms as muscular weakness, painful joints, chronic fatigue, depression, trembling and loss of weight. Whether these symptoms were due to OP insecticides has never been properly resolved. One theory is that the pesticides, which were mainly bought from local firms in the region, might have been contaminated with toxic impurities and it was these impurities that caused the illness.

Others believe that some of the troops might have been affected by sarin, which could have been used by the Iraqis because a thousand rockets loaded with a total of around eight tons of sarin were captured by the allied forces. They were blown up and destroyed and, while this would have eliminated almost all of the nerve agent in the bombs, it would also have scattered some of it, which might have contaminated a wide area.

Sheep farmers were familiar with sheep dipper's flu, which generally started as a headache during the actual dipping and then developed rather like a bad cold. For some, the condition was more prolonged and more serious, with paralysis of the muscles starting about three or four days after exposure. This condition lasted for about a month, and for some people it continued to debilitate them for years.

There are more than forty-two million sheep in the UK, all of which need to be dipped twice a year to control the blowflies and scab which caused widespread infestation in the 1960s and 1970s. Dipping was a legal requirement between the years 1976 and 1992, but now it is up to the farmer whether he wants to protect his flock in this way. Almost from the start some claimed that the organophosphate fluid used in sheep dipping made them ill and that this constituted an industrial disease for which they should get compensation.

The UK Government's Veterinary Medicines Directorate was set up in 1985 to examine such cases and, in the years that followed, they dealt with more than 600 out of a total of 70,000 sheep farmers. This figure is not likely to rise much in future because tough regulations are now in force which require a farmer to sit an examination and be officially registered before he or she can engage in sheep dipping.

One of the most vociferous campaigners against OPs in the UK has been the Countess of Mar, who, in addition to being a sheep farmer, is an independent member of the House of Lords. She was accidentally poisoned by an OP sheep dip in the summer of 1989 and claims she has suffered from its debilitating effects ever since. She describes how its symptoms fluctuate from day to day: some days the brain is unable to think, there can be intense muscle pains, mood swings, speech and writing difficulties, erratic heartbeats and breathing difficulties.

Nor is it only the OPs used in sheep dips that have been a cause for concern. A vigorous, but misdirected, campaign was orchestrated against dichlorvos, more correctly known as dimethyl dichlorovinyl phosphate, which is used to impregnate plastic strips that can then be hung in greenhouses, conservatories and kitchens to kill flies. It is also used in anti-flea collars for cats and dogs. These strips were first put on the market in the 1960s but they were accused of affecting humans, especially those with asthma and babies and particularly when used in confined spaces. They were even banned in Holland in January 1974. In the USA, the EPA investigated them but eventually declared them safe. In fact, the original strips were contaminated with other chemicals which were probably responsible for the adverse effects.

Sometimes OPs can contaminate food, and this is what happened in India on 6 July 1997. The case was reported in the *British Medical Journal* in July 1998 by Rama Chaudhry of the All India Institute of Medical Sciences, New Delhi.

Sixty young men, aged twenty to thirty years, ate their midday meal at a works canteen where they were served with chapatti, cooked vegetables, beans and halva. Over the next three hours all were taken ill with nausea, vomiting and abdominal pains and they were treated at a nearby healthcare centre. Almost all of the men were able to return home that same day, but four were moved to a nearby hospital for emergency treatment when they showed signs of respiratory distress and general muscular weakness. There they were given treatment for their symptoms and three responded and were discharged a week later, but the fourth man died a few days after admission.

To begin with, the doctors thought they were dealing with an outbreak of food poisoning and suspected botulism was the cause. However, none of the food from which the meal had been prepared tested positive for *C botulinum*. Subsequently blood tests on the man who was most seriously ill revealed OPs in his blood, and tests on the chapattis showed they were heavily contaminated with malathion. Once the cause of the illness was known he was injected with the antidotes, atropine and pralidoxime, but it was too late. Malathion rarely gives rise to such adverse reactions, but the incident emphasizes the need for eternal vigilance whenever these insecticides are used.

How safe are OPs? Dr Goran Jamal, of the Institute of Neurological Sciences at Glasgow University, is a member of the UK Government's advisory panel on OPs. In collaboration with the Institute of Occupational Medicine at Edinburgh, he is a leading researcher into the effects of OPs on humans.

According to Jamal, there are three responses to OPs: the acute syndrome, which happens within hours of exposure; the intermediate syndrome, which occurs within days; and the delayed response, referred to as OP-induced delayed neuropathy (OPIDN), which has a time scale of weeks and months. There may even be a fourth – a chronic syndrome – where damage to the nervous system builds up over many years, although this is not easy to link to OP exposure.

OPIDN is hard to diagnose but can be extremely debilitating. It affects the neuropathy target esterase (NTE), an esterase of unknown function which is present at several sites of the body.

The ability to cause degenerative changes in people is associated with the inhibition of NTE and this is now used as a marker to identify harmful OPs. If an OP is shown to inhibit NTE, then it cannot be licensed for use.

△

When chemists undertake research they cannot know what properties the new compounds they develop will have, although today it is possible to guess thanks to our knowledge of the twenty million compounds that have so far been recorded and investigated. Fifty years ago, the accumulated knowledge was insufficient to warn those who dabbled in organophosphorus chemistry what might be in the Pandora's box they were about to open. When they opened the box, though, and found the terrors therein, they were encouraged by warring governments to develop more potent variants. Today we are a little wiser and research is directed towards producing the safest versions of pesticide compounds. The guidelines set by dictators for the investigation of such compounds is, however, another matter.

9. Murder

As the previous chapter showed, phosphorus can be chemically modified to make it one of the most deadly poisons known to humans. Although not in the same league as the nerve gases, the element itself is also a dangerous poison and it does not take much to swamp the body's detoxifying defences. The human body evolved with an ability to detoxify many molecules produced by plants as natural insecticides, but it did not evolve a way of coping with toxic elements, of which phosphorus is a particularly nasty example. It can cope with tiny amounts of phosphorus even on a regular day-to-day basis, levels such as the workers of the match industry were exposed to, or the doses prescribed by doctors, but when the defences are swamped by too large an intake, this toxic element can wreak havoc with vital parts of the body and the liver in particular, the very organ that is trying desperately to get rid of it.

Knowing that phosphorus was poisonous did not prevent people from taking it and, in Victorian times, this often meant eating match heads. It was done either in the belief that it would improve mental ability, or in the hope of committing suicide. The first notion was founded on a long-standing misconception that phosphorus was good for the nerves and brain, the other was more securely based on the knowledge that it was a deadly poison. Young children were accidentally killed by sucking match heads, but when very young babies suffered this same fate it may have been a deliberate act of murder.

A lethal dose of white phosphorus for an adult is around

100 mg, in other words a tenth of a gram,* which was roughly about the amount of phosphorus in a box of lucifer matches. For a child, it may be as little as 20 mg and for a baby as few as ten match heads could be fatal. In one case on record, a twenty-month-old child sucked the heads of a few lucifer matches and, on discovering this, the parents called the doctor, who sent it to hospital for observation. The following day, it fell into a deep sleep, only to wake and begin vomiting; by the fourth day symptoms of jaundice had begun to appear and, on the sixth day, the child died. Other deaths due to phosphorus, where these were known to be accidental, are given in the box opposite.

The smallest dose known to have fatally poisoned someone is 2 mg, the amount contained in a single match head, and that dose killed a two-year-old child. For adults the minimum lethal dose recorded is 50 mg and the largest dose ingested by an adult who survived is 500 mg. This was taken by a woman who tried to commit suicide by swallowing the heads of 500 lucifers.

The time it takes for a lethal dose of phosphorus to kill varies from hours to days. When a large amount of phosphorus is taken into the body, death may occur within hours. A man who committed suicide by taking ten grains of phosphorus (650 mg) died within seven hours, and two children who ate a tin of phosphorus rat poison died within five hours. A large dose of phosphorus immediately attacks the stomach lining, leading to intense pain and the vomiting of blood. A lesser amount will still inflame the stomach wall, but then it is slowly absorbed into the bloodstream, taking between two and six hours, although if the stomach contains fatty food absorption is faster.

As phosphorus-contaminated blood passes through the liver the phosphorus is recognized as alien and removed, but this process of removal causes the element to attack the liver itself. Within a few hours this vital organ can be destroyed and death

---

* An ounce will provide 250 fatal doses.

Accidental deaths from phosphorus ingestion

Whenever phosphorus is accessible to the public there is the risk of deaths from those who use it in ways that were never intended. For example, in Sweden in the nineteenth century women wanting an abortion would eat match heads, although this was often fatal. A most bizarre accident was reported in the *Lancet* in 1890, which told of the death of a young girl who had been smeared with phosphorus rat poison by a fake medium so that she could appear as a glowing spirit during a séance. In Puerto Rico in the 1950s there was a rash of deaths from adults eating rat poison. These people might have believed the myth that elemental phosphorus had brain-enhancing powers, or that it was an aphrodisiac. In the 1960s in Italy there was a craze for eating the contents of fireworks, especially Chinese fire-crackers which contained phosphorus, and this led to several deaths in Bogota.

One of the most unlucky deaths from accidental phosphorus poisoning occurred in the Second World War and was reported in the *British Medical Journal* in 1942. A plane over Germany had been caught in a stream of tracer bullets, one of which hit the plane's navigator in the thigh and passed upwards into his abdomen. When he was admitted to hospital on his return, the doctors noticed that the wound where the bullet had entered gave off a white vapour with the characteristic smell of phosphorus. Although an operation to remove the bullet and its fragments was successful, it was not possible to remove the phosphorus which had already been absorbed by his system. Such bullets contain 3 g of phosphorus and this is sufficient to deliver a fatal dose. For two days the wounded airman appeared to recover, but then his skin grew yellow and he became semi-conscious. Six days later he died and the autopsy showed fatty degeneration of the liver and kidney, which had been caused by the phosphorus.

US soldiers in Vietnam in the 1960s who were injured by phosphorus particles from flares and shells were treated to prevent such secondary poisoning occurring. The phosphorus was dug out of their wounds as quickly as possible and copper sulphate solution liberally applied. This reacts with the phosphorus to form harmless compounds, which can be washed away.

is then inevitable. Even if medical help is available, there is little a doctor can do. The recommended treatment for phosphorus poisoning is to pump out the stomach and wash it with water. The patient is then put on a saline drip containing glucose and calcium gluconate. Remission may occur and the patient will begin to recover, although this recovery may last for only a few days if liver damage is severe, in which case symptoms of the damage, for example jaundice, begin to appear and death will occur within three to four days.

Those poisoned by phosphorus experience a raging thirst due to dehydration, have a noticeable smell of garlic on their breath, suffer diarrhoea and convulsions and pass only a little urine, which is dark coloured and acidic. There is only a fifty–fifty chance of survival. However, remarkable recoveries are on record, such as the man who spooned half a tin of Rodine rat poison into a tumbler of whisky and drank it in a suicide attempt. This gave him a dose of around 300 mg, enough to kill most people. Within two hours he was admitted to Guy's Hospital in London suffering from extreme stomach pains and vomiting; his vomit smelled of phosphorus. He was given emetics for several hours until his vomit no longer smelled of the poison. Three days later, his liver started to enlarge and jaundice set in, but on the sixth day this condition began to right itself and eventually he recovered.

The autopsy of a person who had died from phosphorus poisoning would reveal inflammation and haemorrhage in the stomach and bowel, the liver would show fatty changes and both it and the kidneys would be enlarged, greasy and of a yellow colour. But the most convincing proof of death due to phosphorus exposure would be to turn off all the lights in the mortuary and see its tell-tale glow; even if the glow was not immediately obvious the process of heating specimens of body organs in a flask would drive off the phosphorus so that its glow could be seen.

Phosphorus wasn't the ideal poison with which to dispose of someone. Although it was freely available in the form of medicines, matches and rat poisons, the first of these contained too little to be effective and the second were contaminated with other agents that a potential victim might recognize. So it was the third option that was often the poisoner's choice.

The poisoner who chose phosphorus as his or her agent generally purchased a tin of rat poison, of which Rodine was the most popular brand. Somewhat ironically the introduction of a phosphorus-based rat poison had been welcomed in the nineteenth century as a *safety* measure, in that it replaced arsenic-based poisons. In 1843, the Prussian Government, alarmed at the use of arsenic rat poisons by domestic murderers and suicides, passed a law that only phosphorus could be the poison used to kill such vermin. It was not long before people discovered that these new rat poisons could be just as effective when used for illegal purposes and, in the nineteenth century, such deaths became common in Europe and especially in France.

Indeed, when lucifer matches were no longer available rat poison was the only real source of phosphorus to hand. Even so, phosphorus has a garlicky smell and taste which was difficult, but not impossible, to disguise. Yet phosphorus also had its advantages as a murder weapon because the devastation it produced in the body was consistent with ordinary liver disease, although if this came on suddenly in an otherwise healthy individual there was a risk of phosphorus being discovered at autopsy.

So how did a would-be murderer administer phosphorus to his or her intended victim? Eighteen-year-old Miss Fisher in 1871 sought to dispose of members of her family simply by putting a piece of phosphorus in the family teapot, but the addition of boiling water made its smell very noticeable and no one was harmed – although she ended up in court and was sent to prison. In another incident a woman spread Rodine rat poison on bread and then fried it for her husband's breakfast, but that ruse was also detected and she too went to jail.

In Germany in 1838, a wife prepared some soup for her husband which she laced with phosphorus-based rat poison; she then left it out for him to eat when he came home from work, while she went out to visit relatives. Her attempt to kill him failed because he noticed that his dish of soup glowed when he carried it down a dark hallway from the kitchen to the dining room. It also smelled rather odd and, suspecting that it had been tampered with, he poured some of it into a bottle and took it the following day to the public analyst, who showed it to contain elemental phosphorus. The wife was arrested, charged with attempted murder and ended up in prison.

The toxicity of phosphorus could be much exaggerated, as a report in the Parisian newspaper *La Presse* revealed.* Two men went into a cabaret in the Pas de Calais, Paris, and ordered a cup of coffee. Scarcely had they swallowed the contents than they dropped down dead over the table. The woman who kept the cabaret called the authorities, who immediately suspected foul play and accused her of putting something in their drink. To prove that the coffee was not responsible, she poured a cup for

---

* This is told in R. E. Threlfall's book *100 Years of Phosphorus Making 1851–1951*.

herself, drank it – and also expired in minutes. And the cause? A box of lucifer matches in the coffee-pot. However, phosphorus clearly was *not* responsible for their deaths. (A more likely explanation is that a quick-acting poison, such as cyanide, had been added to the coffee, with the box of matches put there to make it look like an accident.)

Suspecting a death was due to phosphorus poisoning was one thing; proving it to the satisfaction of a court of law was another. The autopsy could suggest that death was due to unnatural causes, and even say that it was probably due to phosphorus, but it was necessary to identify correctly that this was the agent responsible. Various forensic tests for phosphorus were used, and these were named after the analysts who developed them. Scherer's test used filter paper wetted with silver nitrate solution which was placed over the mouth of a flask in which a sample from the body of a victim was heated with dilute acid. If the rising steam caused the filter paper to turn black then phosphorus was present. Another test was devised by Mitscherlich; this required total darkness, so that when a suspect specimen was heated with acid it could be seen to give off a glow of phosphorus vapour. This was condensed under water and then oxidized by nitric acid to convert the phosphorus to phosphoric acid which could easily and accurately be measured.

How much phosphorus was recovered varied considerably. Sometimes it proved impossible to detect any phosphorus – even in the remains of someone who was known to have died from phosphorus poisoning. At the other extreme, in one murder case the Forensic Science Laboratory in Bristol managed to recover 22 mg of elemental phosphorus from a body, one of the highest amounts ever recorded at autopsy.

When phosphorus was present in large amounts, such as in vomited material or the stomach contents of those who died quickly, it could be extracted with the solvent carbon disulphide and its presence noted by putting a drop of this solution on to filter paper. Once the solvent had evaporated, the paper could be examined in the dark and seen to glow. Phosphorus can linger for long periods in the intestines and this was generally regarded as the best place to take specimens for analysis, especially if a body had been exhumed.

The first time this was done, in 1902, it came as a surprise to forensic analysts that they were still able to detect elemental phosphorus in the body of a victim after it had been buried for fifteen months and even to extract it from the remains. In theory there should have been no trace of such a chemically reactive poison after so long, but it depended on the condition of the ground in which the coffin had been buried. If this was water-logged and the coffin air-tight, then it could preserve the poison for a long time, more than three years in one recorded case.

Phosphorus murders

The sale of Rodine rat poison ended in the UK in 1963 when the Animals (Cruel Poisons) Regulations came into force. After that date, cases of phosphorus poisoning in humans virtually ceased. Rodine consisted of a paste of bran and molasses (also known as black treacle) with 2 per cent phosphorus in a finely divided form. A tin of the rat poison weighed one ounce (28 g) and contained ten grains (650 mg) of phosphorus; a teaspoonful would provide a fatal dose.

Rodine was sold in shallow tins, similar to those used for shoe polish, and these were opened and left where rats could find them. These creatures were undeterred by the characteristic odour of phosphorus and, once a rat has eaten something, that substance has to pass through the body (as we have noted) because rats are unable to vomit. It was recommended that a bowl of water be provided near by so that the creature would, when suffering from the extreme thirst that phosphorus could produce, return to it to drink and die.

The ploy in using phosphorus to kill a person, as opposed to a rat, lay in disguising its smell and flavour from the intended victim.

Dr J. Milton Bowers

Whether Dr Bowers disposed of his first wife by poisoning her with phosphorus we shall never know, but he almost certainly disposed of his second wife that way and his third wife was definitely killed with it. As a doctor he had easy access to phosphorus, which was a common medicament, and he could easily have administered it in the guise of a medicine.

Bowers was born in Baltimore in 1843 and married Fannie Hammet, who died in 1865. The second Mrs Bowers was an actress, Theresa Shirk, whom he'd met in New York, and she died in 1881. Within six months of her death Bowers married Cecilia Benhayon-Levy, despite much opposition from her parents, who were alarmed about rumours that he had poisoned Theresa. Their suspicions were well founded. The third Mrs Bowers died in agony in the late summer of 1885, with her body unnaturally swollen, and her family insisted on an autopsy. This showed that Cecilia had succumbed to phosphorus poisoning. The motive seemed clear: Bowers already had a fourth wife in his sights. He was arrested, tried and found guilty although there was a suspicion that his housekeeper, Mrs Dimming, might also have been involved because she had nursed Cecilia while she was ill.

Bowers' conviction seemed less secure when, two years later, Cecilia's brother committed suicide (with cyanide) in a rented apartment and left a letter in which he confessed to murdering his sister. When his death was investigated, it was discovered that the apartment had been rented by Mr John Dimming, the husband of Bowers' housekeeper. The police decided that he had lured Cecilia's brother to the apartment and poisoned him, leaving the confession – which they said was a forgery – by his body.

Accordingly, John Dimming was now tried for murder, but the jury disagreed at his first trial and the second jury acquitted him. By implication this meant that the brother's confession was *not* a forgery – he really had confessed to the murder of Cecilia – and so the verdict of Bowers' first trial had to be set aside. He was sent for retrial, but the case was dismissed.

On his release, Bowers returned to the young lady he'd been seeing prior to his third wife's death and soon married her. They appear to have lived happily together until he died in 1905.

The surprise birthday cake

In 1899, Mary Ann Ansell worked as a maidservant at a house in Coram Street, in the fashionable Bloomsbury area of central London. She was engaged to be married, but the wedding had to be postponed because her intended husband could not afford the 7s 6d for a marriage licence. Resourceful Mary devised a cunning plan to obtain the money: she would take out an insurance policy on the life of her sister Caroline, who was incurably insane and who had been confined in the Leavesden Mental Asylum in Hertfordshire. For a premium of only three pence a week, she insured her sister's life for just over £11, saying that her sister worked at a hospital in the country.

For a few more pence she purchased a tin of rat poison and this she baked into a cake which she sent to the asylum, where it was shared among the inmates of ward 7, although most of it was eaten by Caroline. All became ill, Caroline seriously so, but at the time there were several cases of typhoid among patients in the asylum and so it was a few days before she was properly examined by a doctor. She was then immediately admitted to the infirmary, but died a few hours later. An autopsy revealed the cause of her death to be phosphorus poisoning.

The police were called and were told of the curious circumstances surrounding the cake that had come from her sister. Further enquiries at a shop near Coram Street produced the damning evidence that Mary Ann had indeed bought phosphorus-based rat poison there, shortly before she had sent the cake to Leavesden. The lady of the house where she worked denied any need for such a purchase and so Mary Ann was arrested, brought to trial, found guilty and hanged on 19 July 1899.

Rum and Rodine cocktails

Louisa Highway was born in Wigan in Lancashire, in 1907, one of eight children of a miner and his wife. She left school when she was thirteen and went to work in a cotton mill. At the same time, she left the Methodist church that her family attended and joined the Salvation Army. When she was twenty she became a kitchen-maid at a nearby hospital and at twenty-five she married an ironworker and became Mrs Ellison. In the years that followed she had nine children, although only four survived beyond infancy. Even those were taken into care by the local authority because she neglected them, preferring instead to spend most of her evenings at the local pub. She even spent time in prison during the war for stealing ration books from her neighbours.

Mr Ellison died in 1949 under rather suspicious circumstances – forensic tests could not prove the cause of death and it was eventually put down to kidney failure – and Louisa married a man she had already been spending time with: seventy-eight-year-old Richard Weston, who was recently widowed. He sold his house for £1,600, which joined the £800 he already had in the bank,* and moved in with Louisa, whom he married in January 1950. He died ten weeks later, by which time his money had all been spent.

Louisa next married sixty-seven-year-old Alfred Merrifield, who was retired, in August of 1950. He was the father of ten children and his money, too, was soon gone. In April 1951, they moved to the seaside town of Blackpool, for Louisa to take up a position as housekeeper to an old lady.

It was there, in a modern bungalow named Homestead in Devonshire Road, that cantankerous seventy-nine-year-old Sarah Anne Ricketts lived on a modest annuity that was sufficient for her to employ paid help. She had been twice widowed, both of her husbands committing suicide by gassing themselves, and now she was advertising for a housekeeper. It was on this basis that Mr and Mrs Merrifield took up residence and began to look after her from 12 March. They did it very well.

Mrs Ricketts, who rarely went out, had been neglecting herself, although she was very fond of rum, which was delivered regularly. By the end of that month she was eating better and was so impressed by Mrs Merrifield's care that on 31 March she executed a will in favour of the couple. She would leave them the bungalow, which was estimated to be worth about £3,000 (around £60,000 or $90,000 today). Two weeks later she was dead.

Earlier that month, Mr Merrifield was sent by train to Manchester to buy a tin of Rodine rat poison, which he did at a pharmacist's shop near the station where the assistant remembered him because of his hearing aid. Louisa then experimented with ways in which she might poison her employer. She discovered that a spoonful of Rodine could be stirred into a small glass of rum and, when the contents had been allowed to settle,

---

* The total (£2,400) would be equivalent to a total of £48,000 or $70,000 today.

the liquor in which the phosphorus had dissolved could be poured off, leaving most of the bran from the Rodine behind. The taste of the resulting 'cocktail' betrayed no hint of the lethal dose of phosphorus it contained. Research by a Dr Manning, of the North-Western Forensic Laboratories, showed that 100 mg of phosphorus could be disguised in this manner. Two such drinks would certainly kill.

Louisa probably gave Mrs Ricketts one of these 'cocktails' on Thursday 9 April and, when the old lady started to feel ill, she sent for her doctor, Dr Yule. He found her recovered and, after a cursory examination, went away. Clearly the first dose of phosphorus had not been enough, but at least Mrs Merrifield had established that her employer was not well. A further attempt on her life was made that weekend and Mrs Ricketts was again taken ill. On Monday 13 April, Louisa went to the home of a Dr Wood, who lived a few doors away, saying Mrs Ricketts was seriously ill. He arrived at 6.20 p.m. but could find nothing wrong. However, that night Mrs Ricketts became really ill and, early the next morning, Louisa asked the doctor to come again.

Naturally, he suspected another false alarm and asked his partner to visit the old lady at midday, by which time it was obvious that she was dying. He asked who her regular doctor was and advised sending for him straight away. When Dr Yule arrived two hours later Mrs Ricketts was dead. He refused to issue a death certificate and informed the coroner. Mrs Ricketts' body was immediately removed to the local mortuary, where a post-mortem was carried out and the contents of her stomach and intestines removed for analysis. These eventually revealed traces of bran and phosphorus.

Meanwhile, Mrs Merrifield visited an undertaker and asked about having the body cremated as soon as possible. She made arrangements with the local Salvation Army to play 'Abide with Me' outside the bungalow on the day of the funeral as Mrs Ricketts had also been a life-long Salvationist. It was later thought that it had been this common bond between the two women that persuaded Mrs Ricketts to make a will in Louisa's favour.

Once police enquiries were under way, other incriminating evidence came to light regarding Mrs Merrifield's intentions. In the week leading up to the murder, she had told some of her acquaintances that Mrs Ricketts was already dead and had left her the bungalow. However, the most damning circumstantial evidence was the fact that no tin of Rodine was found and it would have been impossible for Mrs Ricketts herself to have disposed of it, since she could hardly walk and rarely left her bungalow.

Detectives even searched the garden using a metal detector. However, the morning of Mrs Ricketts' death was also the morning that the council dustcart had called and so it could easily have been disposed of. The only evidence that came to light was a dessert spoon in a handbag which Mrs Merrifield had given to a friend the day after the murder, asking her to look after it. This had traces of a sticky material on its surface that could have been rat poison.

It was later estimated that Mrs Ricketts' death had taken place between twelve and eighteen hours after the fatal dose of poison had been given, because the autopsy showed the liver to be pale, enlarged and already undergoing fatty changes. At the trial, the defence maintained that this was due to infective hepatitis. The prosecution evidence was that the victim's stomach smelled of phosphorus when it was opened, traces of bran had been found among its contents, from which 2.5 mg of phosphorus had been extracted at the forensic science laboratories. A further 6 mg was obtained from her intestines.

The trial of both Mr and Mrs Merrifield began at Manchester Crown Court on 20 July 1953 and lasted eleven days. The jury took nearly six hours to find Mrs Merrifield guilty and she was hanged on 18 September. A verdict could not be agreed on Mr Merrifield and a retrial was ordered, but the Attorney-General issued a *nolle prosequi* (unwilling to prosecute) and he was released.

Mrs Ricketts' murder ran true to form for a phosphorus poisoning. She had been given a fatal dose of phosphorus on the Saturday or Sunday, was made ill by it and then recovered temporarily, although she would have been likely to die of liver damage within the next few days in any case. Louisa did not realize this and was unaware that her victim's temporary respite would not last, so she had given her another dose that evening, which precipitated her end – and left the tell-tale traces of phosphorus in her gut.

A cup of tea to start the day, dear?

Mr Davidson married his wife, a native of Burma, when he was stationed in that newly independent colony in 1950, and the couple returned to live in the UK in 1953, when Mrs Davidson became pregnant. For a time they lived with Mr Davidson's brother, who started an affair with his sister-in-law, and it was he who had made her pregnant. Although they planned to keep this from Mr Davidson, he overheard them talking about it. He confronted his brother, who confessed to the affair and its consequences, whereupon he was thrown bodily out of the house. Davidson then vented his anger further with a violent attack on his wife, but he planned a more lasting poisonous revenge and purchased a tin of Rodine.

On the morning of Wednesday 6 April 1955 Davidson made his wife a strong cup of sweet tea which he took up to her in bed and waited while she drank it. He then took the cup downstairs and washed it up, something that struck Mrs Davidson as somewhat unusual. (He repeated this procedure the following morning as well.)

On the Thursday Mrs Davidson began to feel ill with stomach

pains and vomiting and, when her husband returned from work, she complained of a burning sensation in her throat and a strange taste in her mouth. The following day, which was Good Friday, she became very ill, but on the pretext of it being a religious holiday Davidson did not call a doctor and on the Saturday morning it didn't seem necessary, as Mrs Davidson said she felt a lot better. The respite was brief and, later that day, she began to feel worse and asked her husband to go and get a doctor while she went to bed. However, her husband returned from the doctor's house saying he was not at home. Advising his wife to stay in bed and rest, he then went to the cinema and there disposed of the tin of Rodine, where it was found by a cleaner the following day. This seemed such a strange object to find in the cinema that it was taken to the manager's office, where it was discovered to be almost empty.

That evening Mrs Davidson appeared to be getting worse, but she was persuaded that perhaps the pains in her gut were due to constipation and agreed to let her husband give her an enema to relieve it. Relief was only temporary and she was by now clearly very ill. Finally Davidson called the doctor, who arrived at seven o'clock on the Sunday morning, diagnosed a bad case of constipation and asked his colleague to call again on Mrs Davidson during his afternoon round. The colleague found her much worse and she was immediately admitted to hospital.

The story she told the physician who examined her was that she had eaten powdered gramophone record and Davidson confirmed this, hinting that she had hoped to procure an abortion in this way. Stomach washing, however, revealed nothing and in any event the material used to make records was inert and would not have explained her symptoms. That day she got progressively worse and she died the following morning at five o'clock.

An autopsy revealed that she was six weeks pregnant, but death was put down to acute necrosis of the liver which, when microscopically examined, was found to have suffered fatty degeneration. There was no evidence of phosphorus in the body, although there had been a delay of several days between taking samples from the body and their chemical analysis and no special precautions had been taken to prevent any phosphorus that might have been there being oxidized.

Davidson might have escaped arrest had he kept his mouth shut, but after the coroner's inquest he admitted to one of her relatives that he had poisoned her because she had been unfaithful. The relative informed the police, who had also discovered that when Davidson visited the mortuary to identify his wife's body he had twice asked the attendant if there was any evidence that she had been poisoned.

Eventually he was arrested and charged with her murder. All the evidence against him was circumstantial: his purchase of the tin of Rodine, his motive for wanting his wife dead, the discovery of the almost empty tin in the cinema the day after he had been to it, his wife's symptoms, his apparent delays in getting medical help and his remarks to relatives and the mortuary attendant. All were capable of other explanations and his defence lawyers exploited these possibilities, with the result that Mr Davidson was acquitted. Lucky man.

Was Davidson a skilful poisoner? Probably. He may have known enough about the action of phosphorus to use it in small enough doses to destroy her liver in such a way that her death might appear to be due to natural causes a few days later. And, just to make sure that there would be no trace of phosphorus found when she was dead, he persuaded her to have an enema to remove the contents of her bowels, which suggests that he had learned enough from reading about other phosphorus murder cases to know that it was in the intestines that traces of poison could survive long enough to be identified.

Mrs Wilson's husbands

Mrs Mary Wilson, aged sixty-six, was also a skilful phosphorus poisoner who certainly disposed of two husbands with it and may even have got rid of her first husband in that way. Her case is particularly interesting from a forensic point of view, because it provided evidence that elemental phosphorus could linger for more than a year in a corpse. She used a phosphorus-based beetle poison and disguised its taste by mixing it with a popular brand of spicy bottled relish, HP Sauce. Another way that she was suspected of using was a patent cure-all medicine called Chlorodyne, which contained a large amount of chloroform and whose strong odour was enough to mask the smell of phosphorus. This also had the advantage of dissolving phosphorus very easily so that a tablespoonful of the linctus might well contain a fatal dose. The beetle-killer was a paste containing 3 per cent phosphorus, and around a tenth of an ounce of this (around 3 g) would have been enough to deliver a fatal dose of 100 mg.

Ernest George Lawrence Wilson was the third husband of Mary Wilson and he died on 12 November 1957 after a restless night of severe thirst and chest pains. Two weeks later, his body was exhumed in secrecy at midnight and an autopsy carried out at 3 a.m. This revealed no serious organic disease but microscopic examination of his liver revealed extensive and almost complete fatty degeneration. His stomach and intestines were also removed for forensic analysis.

At the same time, the body of Oliver James Leonard, Mrs Wilson's second husband, was exhumed. He had died the previous year on 3 October 1956 and again there seemed to be no evidence of natural disease to account for his death and certainly no evidence of the heart disease he was supposed to have suffered from. However, his liver, like that of Wilson's, was yellow and fatty. It, too, was taken away for further tests, along with Leonard's stomach and intestines.

Dr Ian Barclay of the regional forensic laboratory carried out extensive tests for poisons on the samples from both bodies and concluded that both men had died as a result of phosphorus poisoning. Indeed, the eerie glow of elemental phosphorus was seen not only in the remains of the recently deceased Mr Wilson but, more surprisingly, in those of Mr Leonard, despite his corpse having been underground for more than a year.

Mrs Wilson's trial was held at Leeds Winter Assizes on 24–29 March 1958 and the main thrust of the prosecution's case was the forensic evidence. Dr Barclay said that he had been able to extract 2 mg of phosphorus from Mr Wilson's liver and 0.7 mg from his intestines and had detected its presence in the liver but in an amount too small to measure. In Mr Leonard's stomach there was 0.8 mg and 3 mg in his intestines. There was also wheat bran in both men's stomachs of the type used in the manufacture of the beetle-killer.

Mrs Wilson's defence asked whether Damiana pills might explain the presence of phosphorus in the men's bodies. These were on sale at the time as an over-the-counter nerve tonic, but they also contained strychnine as well as elemental phosphorus. As no strychnine had been detected in the remains, this explanation was ruled out. Another of the prosecution's specialist witnesses, a Dr Price, was called to explain the effect that elemental phosphorus has on the liver. He reported that the fatty degeneration, which characterizes the poison's attack on the liver, was not complete and estimated that the fatal dose had probably been given between twelve and twenty-four hours prior to death.

The jury did not believe that Mr Wilson and Mr Leonard had died from natural causes, despite a vigorous defence that claimed phosphorus had been administered for medical reasons. Mrs Wilson was found guilty and sentenced to death, although this was commuted to life imprisonment.

10. Fortunes from phosphorus

Since the time of its discovery, there has always been a ready market for phosphorus and there have always been people prepared to supply that market. In the early days, of course, it was produced from urine and excrement, but only in tiny amounts – a few pounds a year – and a high price was demanded for it. Things changed with the discovery that bone was a rich source of phosphorus, production increased to several tons per year and the price dropped accordingly. The introduction of phosphorus matches in the nineteenth century boosted demand to hundreds of tons a year and this period saw the emergence of the major phosphorus manufacturers. The use of phosphorus in the world wars of the twentieth century pushed up production to thousands, then tens of thousands, of tons. But it was the peaceful use of phosphorus, to make phosphates for detergents, that sent production soaring in the second half of the twentieth century, and it finally reached a million tons a year.

Phosphorus production in the eighteenth century

Interest in phosphorus in the years following its discovery came mainly from chemists and alchemists. The former were fascinated by it because it proved that beneath the surface of things there was indeed another world; the latter assumed that this wonderful material would hold the key to the philosopher's stone. Neither the chemists nor the alchemists though were able to profit from their research into phosphorus. It was not thought suitable at the time for a gentleman chemist or alchemist to engage in so sordid an activity as *trade*, and so it was left to a humble apprentice, Ambrose Godfrey, to make his fortune from it.

We met Godfrey in chapter 2 as the young assistant who made phosphorus for Robert Boyle. When Boyle had exhausted his interest in the new wonder, Godfrey began to make phosphorus for sale and, for almost fifty years, he had a virtual monopoly on its manufacture. Others tried to make it out of curiosity but few succeeded and it was rumoured that the published methods for doing so omitted some vital piece of information which only Godfrey knew. He never sought to correct this misleading impression, but he was not deliberately secretive about the process and invited curious visitors into his small manufactory to watch it being made. Others such as Kunckel were also making phosphorus in the early years but the quality of Kunckel's phosphorus was generally considered to be inferior to Godfrey's.

Ambrose Godfrey Hanckwitz was born in 1660 at Nienburg in the Anhalt district of Saxony and came to London in 1679 to work for Boyle. The three eventful years that followed were recounted in Chapter 2, years in which he perfected a method of making phosphorus. When Godfrey left Robert Boyle's employment in 1682, he set up his workshop in a field off Maiden Lane, near Covent Garden, and there began to produce phosphorus for sale. It was conveniently placed to exploit the middens of the Bedford House estate which adjoined it, but the workshop needed to be well away from other dwellings because of the awful smells that emanated from it. Godfrey's main employment was at the Apothecaries' Hall, where he eventually became master of the laboratory, but his spare time was spent making phosphorus, and indeed he found it easy to sell all that he made.

His reputation for making the best phosphorus available meant that his business grew, and soon he was employing others and supplying orders from Europe. By 1707, he was a wealthy man and able to afford new premises that had recently been built on the Bedford House site. Godfrey purchased the lease of 31 Southampton Street, leading off the Strand, and there he opened a pharmacy and lived above the shop until his death on 15 January 1741.

According to the terms of the lease, he was forbidden to carry out any obnoxious trade on the premises. This restriction, however, did not apply to the strip of land behind the premises, and it was here that Godfrey built his manufactory, a long and narrow building in which his workmen produced phosphorus. The workshop had glass doors which opened on to the back garden of his house, where Godfrey would entertain customers with demonstrations of what phosphorus could do. A set of engravings was made of the workshop as it existed in 1728 – see Plate 2 – and business continued there for another thirty years, until finally it ran into difficulties.*

The German book collector and connoisseur Zacharias Conrad von Uffenbach visited Godfrey's shop in 1710 and wrote of it:

> We went to the house of the well-known German chemist, Godfrey . . . and saw his incomparably handsome laboratory, which is both neatly and lavishly appointed, being also provided with all manner of curious stoves. We purchased phosphorus at eight shillings a drachm. [A dram was an eighth of an ounce, roughly 3 g.]

In the 1720s Godfrey published several papers on his research into phosphorus and phosphorus acids in the *Philosophical Transactions of the Royal Society* and, as a result of his work, he was elected a fellow of the Royal Society in 1730. He died eleven

---

* The original shop and laboratory were demolished in 1872. A Roman Catholic church, Corpus Christi, now occupies the site where the laboratory stood.

The first automatic fire-extinguisher

Perhaps because the product from which he had made his fortune, phosphorus, was so dangerously flammable, Godfrey had a keen interest in fire prevention. His greatest contribution to this was an automatic fire-extinguisher which worked by exploding a large container of water and so dousing a fire before it could take hold. In April 1723 he demonstrated his device in a purpose-built house in Belsize Park, but it was not entirely successful. A month later he repeated the experiment in Westminster Fields and this time it worked, with the result that Godfrey's 'Fire Watches' became popular and were credited with preventing the spread of fires in London.

---

years later a wealthy merchant and a distinguished member of the foremost group of scientists of the day.

Sadly, Godfrey's business began to falter soon after his death. His eldest son, Boyle, squandered his inheritance, dabbled in alchemy and ended up living on a small pension paid by his brothers, Ambrose and John. They took over the business in 1742, but they fared no better and were declared bankrupt in 1746. The business then passed to Boyle Godfrey's son, who was called Ambrose after his grandfather, and he ran it successfully for the next fifty years.*

Back in the early 1700s Ambrose Godfrey had sold phosphorus wholesale at 50 shillings per ounce and retail at 60 shillings, which at today's prices would be equivalent to roughly £1,000 ($1,500). While phosphorus was not the only part of Godfrey's

---

* It next passed to Boyle's grandson, Ambrose Towers Godfrey, who set up a partnership with his assistant Charles Cooke. The firm of Godfrey & Cooke continued in business until 1915.

business, it must have provided a substantial part of his income, and he is thought to have produced many ounces a week with a yearly output around 50 lb (around 25 kg), which would have earned him around £2,000.

Godfrey had many reasons for keeping his method of making phosphorus secret and, even in 1733 when he finally published an account of his business, he gave away few details. In fact, there was no secret ingredient, as Dr J. H. Hampe, the King's physician, was to discover. He was prompted to ask Godfrey about his ingredients by J. F. Henckel, Counsellor of Mines at Freiberg, whose colleague Andreas Marggraf had found a more effective way of making phosphorus and he wondered if this was the method that Godfrey was already using. Hampe's letter to Henckel, dated 29 August 1735, came to light 200 years later among Henckel's correspondence and it revealed that Godfrey used both urine and excrement as his source of phosphorus but no special ingredient. Marggraf had indeed found a much better way of extracting phosphorus from urine.

Andreas Sigismund Marggraf was born in Berlin in 1709 and became one of the most important German chemists of the eighteenth century. He died in Berlin in 1782, after a lifetime of research, and is best remembered for the discovery that beet could be grown as a source of sugar. Marggraf was also interested in phosphorus and, like those before him, he chose urine as the source from which it could be obtained. But, unlike his predecessors, he was not content simply to repeat the process as reported in the existing literature. He began instead to research the method with a view to improving it.

In 1734 Marggraf found a better way of extracting phosphorus from urine in greater yield by using lead oxide. He reduced urine to a thick syrup and then mixed it with charcoal and red lead (Pb_3O_4).*

---

* In the ratio of 10 pounds of urine syrup, to 1 pound of charcoal, to 3 pounds of red lead.

When this mixture was heated it gave a black powder, which, on continued heating, produced lots of phosphorus. Intrigued by his new method, Marggraf eventually found it best to add red lead to urine to precipitate lead phosphate, which was filtered off, mixed with charcoal and then heated. This caused the carbon of the charcoal to react with the lead phosphate and off came phosphorus. At the end of the reaction the lead had reverted back to lead oxide, which could even be put through the process again.

While Marggraf's work advanced the chemistry of phosphorus by making it easier to produce, it did not greatly add to its availability because it still relied on urine as its source. He explored other possible sources and was able to extract phosphorus from wheat, mustard and common vegetables, but the quantities he obtained were small. Urine remained the material from which phosphorus was made until about 1769, when a far more plentiful source was discovered.

In that year, Carl Scheele and Johan Gahn proved that bone was mainly calcium phosphate. This discovery was to transform the manufacture of phosphorus for the next one hundred years; the smelly days of distilling urine and heating excrement were over.

Carl Wilhelm Scheele (1742–86) was born at Stralsund, Germany, and died aged forty-four at Köping, Sweden, where he had spent most of his life. He started as an apprentice to an apothecary and ended being elected to the Stockholm Royal Academy of Sciences in 1775, after he had discovered chlorine in 1773 and oxygen in 1774.*

Johan Gottlieb Gahn (1745–1818) was born at Voxna, Sweden. He studied at Sweden's Uppsala University, and became one of the leading chemists and mineralogists of his day. The mineral

---

* During his short career he also investigated molybdenum metal, hydrofluoric acid, the green pigment copper arsenide (which became known as Scheele's green) and hydrocyanic acid.

gahnite, which is zinc aluminium oxide, is named after him. Gahn also discovered the first natural phosphate mineral, a green ore found near Beisgau, which was lead phosphate.

In 1769 Gahn and Scheele corresponded with one another about the nature of bone. They knew it contained a compound of calcium, and Scheele wondered whether the other component might be phosphate. However, heating bone ash mixed with charcoal did not release phosphorus even under the strongest heat. Scheele suggested that Gahn should treat some bone ash with sulphuric acid and this he did, releasing the phosphorus as phosphoric acid.* When the phosphoric acid was heated with charcoal, phosphorus poured off in abundance.

When the great French chemist Antoine Lavoisier wrote his *Elements of Chemistry* in 1789, it put the emerging science of chemistry on a sound footing.† He devoted an entire section of the book to the chemistry of phosphorus and how the element could be derived from animal bone. He wrote that phosphorus could form compounds with other elements such as hydrogen, oxygen, nitrogen, sulphur, carbon and some metals.

Lavoisier also reported that charcoal contained a little phosphorus, which made him suspect that the element was more widespread in nature than had previously been supposed. In this, he was echoing the opinion of Marggraf, who had written that the phosphorus salt he obtained from urine was more abundant in urine gathered in summer than in winter. He attributed this to the fact that, in summer, people ate more vegetables and that this could be where phosphorus originated from. Other French chemists found that phosphorus, as

---

* The result of this is to convert the insoluble calcium phosphate of the bone into insoluble calcium sulphate. This chemical reaction releases the phosphate as phosphoric acid, H_3PO_4.

† In this book, he defined for the first time what a chemical element was and named those that he knew, including phosphorus.

phosphate, was present in all kinds of bodily fluids and that some, such as that found in the brain, were combined with fatty matter. These are the phospholipids.

Demand for phosphorus was growing, but neither Scheele nor Gahn was interested in commercial production and the ·manufacture of phosphorus using the new methods began in France, where a chemist, Bertrand Pelletier, started to make it on a large scale in the 1770s. Pelletier was born in 1761 and is remembered for his work on platinum and other metals. He might have become one of France's first chemical entrepreneurs had he not accidentally killed himself by inhaling chlorine gas at the tender age of thirty-six years.

Despite his premature death, Pelletier ensured that France became the centre of phosphorus manufacture in the late eighteenth century and, thanks to his efforts, production at that time was in excess of 200 lb (100 kg) per year. Pelletier reported that he could sometimes produce almost 4 lb from 36 lb of bone ash, which represents a yield of more than 50 per cent. The element could be purified either by squeezing it through chamois leather, or by performing a second distillation, at which point the phosphorus lost its yellow colour and became a white, semi-transparent, wax-like solid. This second distillation could be carried out at 280°C, a temperature much lower than that required to produce the element itself.

Pelletier might easily have died from phosphorus rather than chlorine poisoning. He made his sticks of phosphorus by melting the element in hot water (it becomes liquid at 44°C) then sucking it up into a short glass tube until it 'has risen to within an inch of your mouth' – he then put his finger over the end of the tube to stop it flowing back and transferred the tube to a basin of cold water. The phosphorus solidified and, with a slight shake, it would slip out of the tube and into the water.

Thanks to Pelletier, phosphorus was being made in larger quantities than ever before and it was cheap enough to be used as a source of instant flame, for example phosphorus tapers, as we saw in Chapter 4.

Phosphorus production in the nineteenth century

There was a modest but steady demand for phosphorus in the first three decades of the nineteenth century, when it was used in medicine, for instant light devices and by theatrical entertainers. Some phosphorus was turned into phosphoric acid and thence to phosphates, but the only practical outlet for these was as flame retardants, where their expense would justify the cost (see box).

Flame-proofing with phosphorus compounds

In 1786, it had been suggested by a Mr Arfird that theatre curtains could be fireproofed by dipping them in a solution of ammonium phosphate, but the idea was impracticable because this chemical was not commercially available. In 1821 the respected French chemist Joseph Gay-Lussac took up the idea again and found that dipping hemp and linen in a solution of ammonium phosphate and borax (sodium borate) gave better results. Ammonium phosphate and urea phosphate were also used as flame-retardants for cotton.

Phosphates are still used today in flame retardants for cotton, and these can be chemically bonded to the fabric and so retained through many washes.

Phosphate flame-proofing works as follows: on heating, the phosphate reacts with the cellulose of the cotton to form phosphoric acid, which then catalyses the decomposition of unburned cellulose to a slow-burning char – and the flames die away.

While such uses were welcome, they were not enough to stimulate sufficient demand to support a phosphorus-based chemical industry.

The discovery of congreve/lucifer matches in 1831, though, boosted demand to such an extent that new companies sprang up all over Europe to provide the phosphorus. Manufacture on an industrial scale was begun in Lyon in the 1830s by M. M. Coignet, who improved the Pelletier process and boosted the yield of phosphorus, via a more careful preparation of phosphoric acid from bone. All over France, kilns were built for the calcination of bones. These were self-fuelling, using the gelatin and grease in the bone to keep the kiln going once it had been fired up, and all that was needed was to add more bones to the top of the kiln and rake bone ash from the bottom.

By 1833 the quantity of phosphorus being produced had increased dramatically, and inevitably the price of phosphorus had fallen no less dramatically, to 42s per lb, a fraction of the price it had commanded in the previous century. By 1837 the price had fallen to 21s, as a consequence of fierce competition to supply the match-makers. Export trade in phosphorus was booming, and in the 1840s alone 10,000 lb (4,500 kg) a year was exported from France to England.

However, this state of affairs was not to last for long and, in the 1850s, a new British company named Albright & Wilson appeared on the scene. By 1860 it was able to supply all the phosphorus required for the home market. The following year it too became an exporter and, by the end of the nineteenth century, it was the world's largest phosphorus manufacturer. Arthur Albright and his partner, John Wilson, had succeeded in making their fortunes from this fickle element. As we shall see in Chapter 13, the demise of that company came swiftly and completely 120 years after it was founded, and the unlikely cause of its downfall was a shoal of fish.

Albright was born on 3 March 1811 at the village of Charlbury in Oxfordshire. He was the sixth of ten children born to Rachel Tanner, wife of Wilson Albright. They were a Quaker family, despite the sacrifices this entailed. Only those who accepted the rites of the Anglican Church were able to go to university and without a degree there was no way to enter the legal or medical professions, or even to enter politics, because only Anglicans could sit in Parliament. Consequently Quakers educated their children in Quaker schools and they then went into trade and industry, often with remarkable success – as Albright's career was to demonstrate.

He was apprenticed to a pharmacist in Bristol when he was sixteen and eventually moved to work for John & Edmund Sturge Ltd, a firm of manufacturing chemists in Birmingham. When he was twenty-nine he was taken into partnership with Edmund Sturge, and in 1844 Albright persuaded Edmund that they should start making phosphorus, using bone ash from South America. The demand for phosphorus was growing steadily and, although Sturge Ltd was only a small-scale producer, Albright was convinced that this was the way ahead. He took samples of the company's phosphorus to various exhibitions such as the Great Exhibition of 1851 in London and the Paris Exhibition of 1855.* By this time the price of phosphorus had fallen to 4s per lb, yet Sturge's was still cheaper than that of the leading producer in Europe, the French firm of M. M. Coignet. Soon the price was to fall to less than 3s per lb, but still it was profitable to manufacture.

Around this time Albright and Edmund Sturge dissolved their business partnership, and Albright moved the phosphorus works to Oldbury, near Birmingham. By now he was making around

---

* The catalogue of which reveals the major phosphorus producers of the time: M. M. Coignet of Lyon, Zoeppritz of Freudenstadt and G. A. Reimann of Nuremberg. There were smaller producers in Russia, the US, Bavaria, Belgium, Austria and Sardinia.

26 tons of phosphorus a year and profits in excess of £1,000 a year. At this point John Wilson enters the picture: in 1855 Albright offered him a job and the following year took him into partnership. Wilson had been born in 1834 at Kendal in the English Lake District and was the youngest of ten children. He had the financial acumen to take Albright & Wilson from relative obscurity to its status as a world-class chemical company.*

The production of phosphorus continued to grow and by 1881 had reached 500 tons a year. Quality control was crucial and the phosphorus of Albright & Wilson was regarded as the best available. It was made from phosphoric acid and coke in special retorts, twenty-four of which were stacked in rows inside a furnace, where they were heated for around sixteen hours to drive off all the phosphorus. This was passed through iron pipes to collect as a liquid in a trough of water.

The amount Albright & Wilson could produce in this way was impressive, with 10 tons of bone ash producing a ton of phosphorus; in other words this process released 80 per cent of the phosphorus in the bone. The product was impure but they devised a method of purification which involved stirring molten phosphorus under a solution of acidified potassium dichromate for an hour or so. This oxidized the impurities, leaving a colourless liquid which was cast into sticks and stored under water.

By 1863, there were twenty-seven furnaces at the Albright & Wilson works in Oldbury which were filled with a total of 648 retorts. Their factory was surrounded by a high wall and with the exception of employees no one was allowed on to the site, which gave rise to rumours that they had some secret process by

---

* John Wilson (1834–1907) also came from a Quaker family. He and Arthur Albright married two sisters, Catherine and Rachel Stacey, daughters of another Quaker, Samuel Stacey, and his wife Mary. John and Catherine Wilson had two children; Arthur and Rachel Albright had five.

which they were able to produce phosphorus more cheaply than their competitors.* Bone continued to be their raw material, but when this became increasingly hard to obtain from 1870 onwards they started importing the mineral sombrerite, which had been discovered in 1862 on the West Indian island of Sombrero. Because this phosphate was of guano origin, deposited as bird droppings over millions of years, it was soft and easy to grind and free of the troublesome fluoride which was present in almost all the phosphate minerals that were known at the time.

The market in phosphorus was highly competitive, but Albright & Wilson had the benefit of easy access to a cheaper source of energy than their competitors. Coal prices in England were lower than those in France and Germany and that was the secret of their success. But even this advantage was to be strengthened in 1880 when they started using synthesis gas[†] to heat the furnaces and phosphorus yields improved further. That year their output reached one million lb of phosphorus (450 tons; 450,000 kg) but the price had fallen below 2s per lb and was in the 1890s eventually to drop below 1s per lb, but by then there had been a revolution in the way phosphorus could be extracted directly from its ores.

Back in 1861, A. Muller had patented a method in which mineral phosphate, sand and coke (carbon) were ground together and heated strongly – and, of course, off came the phosphorus. He had proved that, in theory, there was no need to convert the phosphate to phosphoric acid before heating it with a source of

---

* The company had trouble with highwaymen waylaying the firm's wages – a problem that was solved in 1865 when Albright persuaded a fellow Quaker, Herbert Lloyd, to open a pay office in Oldbury. This was the start of Lloyds Bank, which still has a branch in Oldbury.

† Made by blowing steam and air through red-hot low-grade coal. This gives a mixture of hydrogen and carbon monoxide gases, which burns to give higher temperatures than burning even the best-grade coal.

carbon. However, the reaction only worked at a commercially useful rate if very high temperatures were achieved, and this question of temperature presented insuperable problems for the furnace technology of the day.

The answer was to heat the mixture in an electric furnace, and in 1888 two patents were filed for making phosphorus by the Muller method. J. B. Readman filed his patent on 18 October that year, while T. Parker and A. E. Robinson, of the Electric Construction Company of Wolverhampton, filed theirs seven weeks later on 5 December. When they realized that they had been beaten by Readman, they negotiated to buy his patent and immediately started commercial production of phosphorus using the new process.

Albright & Wilson were quick to see that this was where the future lay, thanks to George Albright, one of Arthur's sons who had entered the business and was head of research at Oldbury. He had been educated at Cambridge, where he had studied chemistry and he recognized the benefits which the new process offered, describing it as one of 'brutal simplicity'. Albright & Wilson took over the Electric Construction Company in 1889 and installed electric furnaces at Oldbury. Patents were also taken out in other industrial countries around the world. So successful was the new process that production of phosphorus by the retort method ceased in 1895.

The new phosphorus plants had to be near a cheap supply of electricity, which generally meant hydroelectricity. (It requires about 14 megawatt hours of electricity to produce a ton of phosphorus.) As a consequence Albright & Wilson looked to North America and, in 1897, they built a massive plant near Niagara Falls, but it was the need for cheap electricity which eventually contributed to their downfall sixty years later. Meanwhile the demand for phosphorus grew, as other uses emerged, and after the Second World War more than 90 per cent of the phosphorus produced was turned into phosphoric acid, from

The chemistry behind the manufacture of phosphorus

This has not changed in 300 years and it consists of heating phosphate (PO_4^{3-}) and carbon (C). The carbon removes the oxygens, forming carbon monoxide (CO), and this gas carries the phosphorus off with it, which is condensed in water.

Over the three centuries following its discovery the source of phosphate changed from being human and animal excreta to animal bone, and finally to mineral rock. The source of carbon changed from charred organic material to charcoal, and finally to coke. The method of heating went from being a charcoal furnace to coal, to synthesis gas, to electricity. Materials and technology changed, but the chemistry remained the same.

The act of passing a high density of electric current through a furnace charged with pellets of phosphate ore, coke and sand causes the chemical reaction between them to occur. The temperature in such a furnace reaches 1,400°C. The process can be continuous, with pellets being added continually to the top of the furnace. A mixture of carbon monoxide and phosphorus vapour evolves, while the calcium component of the rock, and other metal impurities, reacts with the sand to form a molten slag, which can be tapped off from the base of the furnace. Even this is a saleable commodity as it can be used for surfacing roads.

which phosphates were made. These were to become key industrial chemicals, especially high-grade phosphates, which are used in the food industry for reasons explained in Chapter 12.

Phosphoric acid also had a market in metal treatments. In 1869 G. Rose found that dipping iron into phosphoric acid made it rust-proof. It does so by forming an impervious layer of insoluble iron phosphate on the surface which protects it against reaction with oxygen and water, the twin causes of rusting. The first use of this new metal treatment was to produce rustless stays for corsets.* Around this time C. A. Wurtz had discovered that it was possible to plate other metals with nickel by heating them in a bath of nickel dihydrogen phosphate at 100°C and that this provided an alternative to the electroplating method for producing such articles.

Phosphorus production in the twentieth century

At the start of the twentieth century there were three companies manufacturing phosphorus on a large scale, all using the electric-furnace process: Albright & Wilson in Oldbury, Birmingham, M. M. Coignet near Lyon and Chemische Fabrik Griesheim-Elektron at Frankfurt-am-Main. Phosphorus production, though small by modern standards, nevertheless kept on rising and had reached 1,200 tons by 1910. The forthcoming world war saw production in the UK more than double to 2,500 tons by 1918, and this high level of production was reached again during the Second World War. However, by then the US had become the major producer of phosphorus, with a turnover of 38,000 tons in 1939.

After the Second World War phosphorus production took off, rising to 400,000 tons per year in the US alone by 1962, which accounted for half the world output. By 1976, the production of phosphorus worldwide topped one million tons. As we shall see in Chapter 12, the washing-machine detergent market created a demand for sodium tripolyphosphate (STPP), which

---

* Phosphoric acid is still sold in DIY shops as a rust-remover for iron objects, such as ornamental railings, wrought-iron work, car-bodies and so on.

was originally manufactured from phosphorus. And, while producing this was not cheap, it was easily justified because a ton of phosphorus generates 2 tons of phosphorus pentoxide when it is burned and this reacts with water to produce 3 tons of phosphoric acid, which ultimately yields 4 tons of STPP. The use of this chemical in detergents made enormous profits for phosphorus producers.

A modern phosphorus-making furnace is about 12 metres across and works at around 500 volts and 60 megawatts, producing around 30,000 tons a year. It is heated by three giant electrodes made of carbon and fed with nodules of pre-treated phosphate rock, coke and sand. The phosphorus that is given off is passed into a spray of warm water and condenses as a liquid, which is how it is shipped, in heated tankers or rail cars.

Most phosphorus is burned to form the oxide (and then converted to the acid by dissolving it in water), but around 10 per cent goes into making other chemicals, such as phosphorus trichloride (PCl_3), which is employed to make flame-retardants, pharmaceuticals and insecticides, or phosphorus sulphides such as P_4S_{10}, which is used to make lubricant additives, or red phosphorus, which is used in matches, fireworks and as flame-retardants in plastics. Some phosphorus (red) is heated with zinc to form zinc phosphide (Zn_3P_2), which is no longer used as a medicament (see Chapter 3) but is sometimes used as a rat poison, and some phosphorus is heated with magnesium to form magnesium phosphide (Mg_3P_2), which is employed to make self-igniting distress flares (see Chapter 14).

By the mid-1980s the demand for elemental phosphorus was starting to decline as the introduction of newer technology made it possible to convert phosphate rock directly into phosphoric acid of a purity sufficient to make detergent and food-grade phosphates. By now the demand for STPP from phosphate-manufacturers had declined sharply in the face of criticism from environmentalists, who protested that phosphates were polluting lakes and rivers. By the 1990s, demand had fallen to such an extent that old phosphorus plants were phased out and phosphoric acid plants were built instead.

Although the amount of elemental phosphorus produced in developed countries continues to decline, global production is still in excess of half a million tons a year. The continued demand for elemental phosphorus is guaranteed by the need for phosphorus trichloride (PCl_3), the starting material for the world's top-selling herbicide, glyphosate, which is made by Monsanto and marketed as Round-up.*

The 1990s saw even greater changes in phosphorus manufacture and at the end of the twentieth century there were only two phosphorus plants in the US: at Soda Springs, Idaho, and Pocatello, in the same state, which is the world's largest. In Europe there is only one phosphorus plant, at Vlissingen in the Netherlands.

Phosphorus is still a profitable industry, although not the cash-generator it once was, and the fate of Albright & Wilson is a sad example of how good intentions, good economic reasons and a strict adherence to safety regulations can still result in a massive disaster. It has always been necessary to take great care when making phosphorus – its highly flammable nature means that the highest standards of manufacture, storage and transport must be maintained at all times, and there have been relatively few accidents at plants where it is produced. Nevertheless, some notable disasters have happened once it has left the factory, and these are described in the next chapter.

---

* The phosphorus trichloride reacts with water to form phosphorous acid (H_3PO_3). This chemical reaction is the only way in which this acid can be made on a commercial scale.

11. Unlucky days

The two most dangerous properties of phosphorus are its toxicity and its flammability. Whereas toxicity has historically been an industrial problem, as it was in the days of phosphorus matches, it is the flammability of phosphorus that has led to the more spectacular accidents.

The nature of phosphorus manufacture is such that it tends to be made in locations far from where it is to be used, which means it has to be transported in tonnage quantities. The potential for calamity therefore exists, and this potential has been realized in the form of spectacular accidents resulting in appalling fires. Two of these have occurred in the USA. There, phosphorus is usually shipped in specially designed rail cars each capable of holding as much as 60,000 litres (16,000 gallons, 110 tons) of liquid phosphorus. The tanks are equipped with heating coils to keep the phosphorus molten, and this is protected by a layer of water and then a further layer of the inert gas, carbon dioxide, which fills the remaining space. The rail cars have an inner steel shell 11 mm thick, able to withstand a pressure of around 35 bar (thirty-five times atmospheric pressure) and an outer steel shell 3 mm thick with 10 cm of glass-wool insulation between the two shells. These safety precautions are intended to provide complete protection even if a tank comes off the rails and rolls over – provided both the inner and outer steel shells are not ruptured. That is never likely to happen ... except, perhaps, on April Fool's Day.

Brownston, Nebraska, 1 April 1978

A train comprising eighty-five loaded freight cars was being pulled by three locomotives and was travelling at 100 k.p.h. (60 m.p.h.). Among the freight was a full tank of molten phosphorus. At 4.45 a.m. the train had just passed through the sleeping town of Brownston when part of a connecting bar worked loose and started to drag along the track, catching a switch and forcing the wheels of one of the cars to leave the track. The following cars then began to pile into the derailed car and these were loaded with timber, potatoes, frozen fries, dried fruit, paper bags – and liquid phosphorus. The tank car was the ninth to derail and it was soon buried upside down beneath other cars. As it derailed, the phosphorus tank was slightly damaged by the wheel of another car, which cut a gash an inch wide and about a foot long (2.5 cm by 30 cm) penetrating both the outer and inner steel shells. Phosphorus began to leak out and soon ignited. The fire it caused began to burn fiercely beneath the overturned tank.

Fire units were on the scene within fifteen minutes, saw what was happening and contacted the Chemical Transportation Emergency Center in Washington DC to ask for advice. They were told to deluge the phosphorus fire with water or, if this was not possible, to bury it with sand or, failing either of these options, to let the fire burn itself out. As there was no water supply near by, nor sand, the only option was to let the phosphorus burn. By now, the flames were soaring skyward, the heat was intense and burning molten phosphorus was running out of the wreckage. Bulldozers were positioned to try to stem the flow. So intense was the heat that the men operating them had to wear special insulated suits, but nevertheless they managed to build a dyke over a metre high to contain the flood of phosphorus. By 11.30 a.m. the crisis seemed to be over, though the phosphorus fire was burning steadily. Eight minutes later the phosphorus tank exploded. An estimated 50,000 litres of phosphorus and debris were scattered over a wide area and the whole engulfed in a sea of flame. Twelve hectares (30 acres) of crops in the surrounding fields were set alight. Luckily only six of the 150 firefighters who were in the vicinity were injured, although two of them were badly burned.

After that, the fire was left to burn itself out, which took a further three days. Thankfully the derailment had not happened in a built-up area, but the next accident was not so fortunate in this respect.

Miamisburg, Ohio, 8 July 1988

On a hot July afternoon in 1988, a forty-four-car train was travelling through Ohio when it approached the small town of Miamisburg, 14 kilometres (9 miles) south of Dayton. Indeed, the weather was so hot that the rail track had developed a 'sun kink', a misalignment caused by overheating. Part of the freight was a tank containing 50,000 litres of molten phosphorus. This was *en route* from the Albright & Wilson phosphorus plant at Varennes in Quebec, Canada, to Fernald, Ohio. As the train travelled over the heated track, the phosphorus tanker was one of seven cars that failed to negotiate the kink and it was derailed. Not only that, it rolled over and one of the welded metal sections of the tank failed, releasing molten phosphorus. This quickly caught fire, and burned fiercely beneath the derailed car, with the result that the whole tank threatened to become another giant phosphorus bomb.

The emergency services were quick to react. The 15,000 residents of Miamisburg were speedily evacuated from their homes, along with 40,000 others in surrounding neighbourhoods. They watched from a safe distance as the cloud of dense white smoke poured skywards from the burning phosphorus.

Meanwhile, local fire teams arrived at the scene and began to spray the leaking tanker with water. The salvage experts were worried about a tank of sulphur which had also been derailed, concerned that this too might ignite.

A decision was taken to rescue this tank and, to facilitate the rescue, firemen agreed to reduce the streams of water they were playing over the burning phosphorus. As recovery crews tried to drag the sulphur car away, they disturbed the phosphorus tank, which suddenly exploded. A mass of burning phosphorus was released and many people engaged in the rescue operation were injured: almost 200 required medical treatment, mainly for skin, eye and lung irritations, but none was seriously burned. The decision was then taken to allow all the phosphorus to burn off, which took four days.

Portishead, August 1990

Albright & Wilson have a specially constructed chemical storage depot at Portishead, near Bristol, and this contained more than 30 tons of phosphorus stored in 166 drums, each holding 200 kg. In the first week of August 1990 two of these drums mysteriously caught fire, but the blaze was quickly dealt with by the local fire brigade and the summer heatwave was blamed. That month saw temperatures in southern England exceed 35°C, and this may well have caused the disaster that was to happen two weeks later, when another drum started to burn in the middle of a hot night. Within minutes, the fire had spread to other drums and, this time, there was no saving the depot as drums surrounded by molten burning phosphorus burst open, adding more fuel to the fire.

More than 100 firefighters were eventually called in before the blaze was brought under control, and there were no casualties. A thick plume of white smoke blanketed the area and drifted for 24 kilometres (15 miles) down the coast to the holiday resort of Weston-super-Mare, where it set off smoke detectors in many of the hotels.

Placentia Bay, Newfoundland, 1968

The most serious accident involving phosphorus was not a fire but an environmental disaster: this was the Placentia Bay crisis of 1969. It later came to be referred to as 'the red-herring affair', an apt name for a variety of reasons. The first act of the tragedy took place in December the previous year on Friday 13 December, traditionally a day of ill omen. The drama played on for around six months, during which time tens of millions of fish died.

It had seemed a good idea to build the world's most up-to-date phosphorus plant at Long Harbour, a cove in southern Newfoundland which fed into Placentia Bay. This area of great natural beauty was home to a community of 275 fishermen who made their living catching herring, cod and lobsters. They were worried when plans were announced to build the new plant but were told that care would be taken to ensure their livelihoods would not be affected and that, in any case, the new plant would create about 375 jobs for local people.

Albright & Wilson decided to build the $40 million* phosphorus plant in this remote location for sound economic reasons. In addition to the benefits of low-cost hydroelectricity, from a power plant at Bay d'Espoir, there were transport costs to consider. To make phosphorus you need phosphate ore, coke and sand in large quantities. Sand was available locally, coke was to be brought from the UK and good-quality phosphate ore was to be shipped from Florida. The location of the plant at

---

* Equivalent to more than $500 million at today's prices.

Long Harbour minimized the transport of raw materials and the product could be shipped from there because the deep-water harbour remained ice-free all year round. The phosphorus was to be shipped to where it was needed in two specially constructed vessels which were capable of carrying millions of litres of liquid phosphorus across the Atlantic to Europe. In 1966, construction began and, on Monday 8 December 1968, the plant opened.

Five days later, on Friday the 13th, all the fish in the harbour were floating on the surface, dead. The plant was obviously to blame and it was immediately closed down. What could they have discharged into the bay to poison all marine life? In fact, they were disposing of large quantities of waste. Each day the plant was in operation it released 10 tons of fluoride, 3 tons of sulphur dioxide and 40 kg of cyanide, all dissolved in water. In addition about 500 kg of phosphorus were carried away from the plant in waste water. This came from the vessels where the phosphorus vapour from the electric furnaces was cooled by a spray of water which caused it to condense as liquid phosphorus.

The fluoride came from the phosphate rock, which is fluoro-apatite, $Ca_5(PO_4)_3F$, and this was removed when the pellets of phosphate were made for the electric furnace. The sulphur dioxide and cyanide came from the coke-drying. All these processes took water from a local creek, and the waste waters were then collected in a large pool where they were diluted with more water. The waste was not immediately discharged into the sea but kept for a few days in a retaining pond, during which time it was assumed that the phosphorus it contained would be oxidized and hydrolysed and rendered safe. Water from this pond was finally diluted with an equal volume of sea water, which turned it cloudy as the fluoride formed insoluble calcium and magnesium fluorides, and then it was pumped out to sea.

The people at the plant reviewed the various operations and decided that the toxic waste water had not been properly treated before being disposed of. Before the plant started up again the company placed lobsters in pots all round the harbour and waste outlets and thereafter monitored them every day. When the plant resumed production these crustaceans were fit and healthy, and they remained so in the weeks and months ahead.

Then, on 2 February 1969, a massive shoal of herring were found floating dead in Placentia Bay, and the dead fish had a curious red discoloration around their gills which gave rise to the 'red herring' tag by which the affair became known. What had killed them? Some suggested they had died as a result of bacterial infection, but there was nothing to indicate the nature of this infection. Perhaps they had swum through waters infested with a bloom of red algae which can release toxins that are deadly to fish and which can wipe out whole shoals. Some fishermen, remembering what had happened two months earlier in Long Harbour cove, blamed the phosphorus works, but the managers there pointed out that the lobsters in pots were thriving. To reach the bay any toxins would have had to pass by them.

Two days later, another shoal was wiped out near Fair Haven, further along the coast of Placentia Bay. Five days later, the same thing happened at Arnold's Cove 40 kilometres (25 miles) from Long Harbour, then at Weedy Island on 11 February, at David Cove on the 14th, and finally a massive fish kill of an estimated ten million herring floated to the surface near Jude Island. Each disaster was further from Long Harbour than the previous one but each reinforced the fishermen's belief that the phosphorus plant must be the source of the toxin. The prevailing current around Placentia Bay first passed Long Harbour before moving on to the other locations where kills happened, and the time between the kills corresponded to the rate at which the current moved around the bay. Yet the lobsters were still alive and kicking.

In March, the death toll among the fish seemed to abate, but then, in April, came the worst fish kill so far and, on 2 May, the plant was forced to shut down. This did not immediately stop the fish kills because, as the second batch of contaminated water continued its journey around the bay to the ocean, it engulfed fresh shoals of herring and wiped them out.

The closure of the plant was a commercial disaster for Albright & Wilson. Unable to meet debt payments, they were on the verge of bankruptcy and were saved only when a US company, Tenneco, purchased a 49 per cent stake in the firm. Tenneco agreed to fund the construction of effluent ponds into which all waste water would be pumped and then treated with lime (calcium oxide) to precipitate the fluoride and neutralize the acid. After further treatment, the water was then reused within the plant. There would be no more discharges of any kind of waste to the sea.

On 11 July, the plant reopened and the new system of water recycling worked. The plant was saved. In 1978 Tenneco purchased the remaining 51 per cent of the company.* The Long Harbour plant finally began to produce phosphorus and ship it across the ocean to the UK, but the boom years of manufacturing elemental phosphorus were coming to an end in any case and, in 1989, the Long Harbour plant was closed down for good.

So what had killed all those tens of millions of fish? The answer was phosphorus. The dead fish that were analysed showed the presence of phosphorus in their bodies and although the levels of the element in sea water in Placentia Bay were incredibly small, research was eventually to show that it was indeed the culprit. Phosphorus in sea water can be toxic to fish at unbelievably low levels.

The Fisheries Research Board of Canada published a report into the incident in 1972 which included work by G. L. Fletcher,

---

* The rescued Albright & Wilson continued in business as a separate company and its shares were eventually sold by Tenneco in the early 1990s. The company was then taken over by Rhodia in 1999.

of the Marine Sciences Research Laboratory of the University of Newfoundland, who had carried out intensive studies into the effect of phosphorus on marine creatures. In a landmark paper he reported that when cod were put into sea water containing only one part per billion of elemental phosphorus they absorbed it into their bodies at an alarming rate. Not only that but they concentrated it in their liver to such an extent that, after only eighteen hours, it reached a level 25,000 times higher than that in the surrounding water, and that level was enough to kill them. Some marine species, like lobsters, were immune to phosphorus at these low concentrations, which rendered them useless as environmental monitors. Even though the phosphorus in Placentia Bay was diluted to almost immeasurably low levels, a shoal of fish swimming in such polluted water for a day or two would all die.

Against all expectations phosphorus had shown that it was able to persist in an environment that should have been able to oxidize and hydrolyse it safely to phosphorous and phosphoric acid. Solid or liquid phosphorus will react with oxygen and as it does so it will generate enough heat to boost the rate of reaction to such an extent that it bursts spectacularly into flames. However, phosphorus that is dispersed in water is very slow to oxidize, and those who designed the plant at Long Harbour did not appreciate just how long this could take. Without the knowledge of how sensitive fish are to this toxic element they inadvertently added yet another chapter to the shocking history of the element.

△

There is no getting away from the fact that the manufacture and transport of phosphorus is always going to be fraught with danger. That is the nature of the beast. The incidents described in this chapter illustrate the problem and show that, despite the most elaborate safety precautions, accidents can happen. What makes phosphorus particularly risky is its triple threat of (1) spontaneous flammability in air; (2) causing intense fires when it burns; and (3) being a highly toxic material. Because phosphorus can only be moved in bulk as a warm liquid, the first of these threats has the potential always of turning an accident into a disaster. The price of phosphorus has to be eternal vigilance.

Yet even when phosphorus had been converted to phosphates, which are unreactive, non-toxic, clean, white powders that are benign enough to be added to foods or used in washing powders, there still lurked an unsuspected calamity of potentially global proportions. Before we look at the threat that phosphates posed, we need to understand the remarkable role that Nature has devolved on to this element, in effect giving it the power to regulate all life on Earth.

12. The supreme ruler

> Life can multiply until all the phosphorus has gone and then
> there is an inexorable halt which nothing can prevent.

The above quotation is taken from Isaac Asimov's book *Asimov on Chemistry*, and is a succinct way of expressing Liebig's Law of the Minimum, which states that a species responds only to the nutrient that is in shortest supply and this is the limiting factor on the growth of a given population. If its amount increases, so will the population, whereas increasing the amount of other, non-limiting nutrients will have no effect. In a chapter entitled 'Life's Bottleneck', Asimov explains why phosphorus is *the* limiting factor to life, both on land and in the sea.*

It took humans a long time to realize the importance of phosphorus to their planetary well-being. When farming first began after the last Ice Age, farmers unknowingly but consistently reduced the amount of phosphorus in the soil by cultivating the same plot of ground year after year, with the result that the more they farmed, the greater the loss of this vital element from the soil. To a limited extent they could replace some of the lost nutrient by spreading manure and compost. Today we replenish

---

* Asimov was very good at explaining limiting factors in other areas too. For example, in his book *The Relativity of Wrong*, published in 1989, he explains how the limiting factor of fuel capacity rules out interstellar travel, so we can never journey to the stars – which also explains why we can never be visited by UFOs either.

it completely with fertilizers, but we only adopted this practice in the nineteenth century, and in earlier ages human ignorance doomed many societies to malnutrition and occasionally to even greater disasters.

King Edward III ruled England for fifty years from 1327 to 1377. During the middle years of his reign the population was ravaged by the Black Death and around one and a half million people died, out of a population of five million. One theory as to why the plague was so devastating points the finger at phosphorus, or rather the lack of it, in the soil. Without enough phosphorus, crop yields had been declining in the years leading up to the plague and, by the time the disease appeared in England, there was already widespread malnutrition, leaving the population far more vulnerable than they might otherwise have been.

How do we know this? The answer is to be found in the archives of Oxford University. In the Middle Ages, the tenant farmers of the village of Cuxham, 16 kilometres (10 miles) south-east of Oxford, were obliged every year to send to their landlord, the University of Oxford, exact details of the crops they had grown, and there the records have remained to this day, carefully written out in Latin. What they reveal is a disturbing picture of decline, as must inevitably happen with traditional methods of farming and despite the fact that the villagers of Cuxham used the three-field system in which crops were rotated. One field grew wheat, for human consumption, one grew a crop as animal fodder, such as oats, while the third lay fallow, this being necessary to regenerate the fertility of the soil, which to a certain extent it did.

Ed Newman, an ecologist at the University of Bristol, has made a careful analysis of the crops grown by the peasant farmers of Cuxham in the years 1320 to 1340 and has shown convincingly that the fertility of the land was in decline. Newman's analysis of the nutrient content of the crops that were harvested shows that they were removing more phosphorus from the soil than the farmers and Nature could replenish. Poor nutrition would have been the result, and the picture was probably the same all over Europe.

If Newman's analysis is correct, it would explain why the population of England stagnated at five million in the early fourteenth century and why it was poorly equipped to cope with the Black Death. When the plague abated, in 1351, the reduced population could once again be adequately nourished by the three-field system of farming. But, perhaps not surprisingly, it was to be another three centuries before it again reached five million. Today the population of the UK is fifty-five million and, despite being a relatively small island, British farmers could easily produce enough food to feed all of them.*

How could land which once struggled to support five million souls now feed an extra fifty million people? To answer that question, we need to look at phosphorus from a global perspective.

The phosphorus cycle

There are five elements that make up DNA and they are absolutely essential to life: carbon, hydrogen, oxygen, nitrogen and phosphorus. Phosphorus is also the key component of the workhorse molecule ATP (short for adenosine triphosphate), which provides the energy to keep cells alive and active. But phosphorus plays many other roles in the body too: it is part of the lipids that make up cell membranes; it is a key component

---

* The UK currently produces about 80 per cent of the food it needs, preferring instead to increase the land devoted to trees and to import foods, such as rice, olives, wines and tropical fruits, which are in demand but which its climate is unsuited to growing.

Phosphorus in living things

It is the convention when talking about phosphates in nature to refer to all of the different forms under the collective title of 'phosphorus', rather than to distinguish each form by its strict chemical name. All phosphorus in the natural world is present as one form of phosphate or another (see Introduction).

In organisms, phosphorus exists as phosphate esters, in which one or two organic groups are attached to the four oxygens that surround the phosphorus. For example, DNA has two carbohydrate (ribose) groups attached. In ATP, there is a chain of three phosphates with an organic group, adenosine, attached to one end. This curious arrangement confers upon the molecule a remarkable chemical reactivity which enables it to drive processes that would otherwise not take place.

---

of molecules that carry messages around the body; and in mammals it is the basis of the skeleton.

Clearly phosphorus in some form or other must be part of our diet, which means it must come either from plants or from other animals. These, in their turn, may be part of a food chain reaching back to the inorganic forms of the nutrient element at the start of the chain. In the case of phosphorus, this form is phosphate (see box). What makes phosphorus different from carbon, hydrogen, oxygen and nitrogen is the fact that it cannot circulate freely about the planet as part of the atmosphere. Because phosphate cannot fly, its movements are restricted and that which does move is carried from the land to the sea by streams and rivers. A little phosphate does get into the atmosphere as dust or sea spray, but the amount is trivial and cannot remedy the endless drain of this element from the land. Some phosphate is transferred back from the sea to the land by fish-eating birds whose droppings have built up sizeable deposits of phosphate on Pacific coastal regions and islands.

Phosphorus circulates through the environment in three natural cycles. The first of these is the inorganic cycle, which refers to phosphorus in the crust of the Earth, of which there is more than a *billion billion* tons. Around thirteen million tons of this is released to soil each year and similar amounts pass into rivers, lakes and seas and are deposited as sediments. Through the aeons, phosphorus has moved slowly through the inorganic cycle, starting with the rocks which slowly weather to form soil, from which the phosphate is gradually leached from the land into rivers and onward to the sea, where it eventually forms insoluble calcium phosphate and sinks to the sea floor as sediment. There it remains until it is converted to new, so-called sedimentary rocks as a result of geological pressure. In the course of millions of years these rocks may find themselves uplifted to form new dry land – and the cycle begins again.

Some deposits accumulated billions of tons of phosphate, such as the Phosphoria deposit of the western USA and that in Morocco. Large deposits are also to be found in Australia, Florida, China and many other countries. The deposits of Christmas Island, in the Pacific, are of guano origin and are fast being used up, but other reserves of phosphate ore will last a thousand years or more.

Imposed on the inorganic cycle are two *organic* cycles which move phosphorus through living organisms as part of the food chain: a land-based phosphorus cycle which transfers it from soil to plants, to animals and back to soil again; and a water-based organic cycle which circulates it among the creatures living in rivers, lakes and seas. Whereas the primary inorganic cycle takes millions of years to complete a revolution, the land-based cycle takes a year and the water-based organic cycle only weeks. It is the amount of phosphorus in these two cycles that governs the mass of living forms, the so-called biosphere, which land and sea can sustain.*

The amount of phosphorus in the world's soils is roughly 150 billion tons and a similar amount is found in the oceans. The presence of so much phosphorus does not guarantee that the element won't be in short supply, any more than the fact that the Earth's surface is mainly covered with water guarantees that there won't be deserts on land. There are phosphate 'deserts' over large tracts of the oceans.

Cultivated land can easily be depleted of its phosphorus, and this was a particular problem when primitive farming methods were used. These speeded up the removal of phosphorus from the land, and the ensuing depletion may well have been responsible for the disappearance of earlier civilizations. The phenomenon of some abandoned cities may well have been caused by a steadily dwindling food supply from land that gradually produced less and less. Even provinces of the Roman Empire are known to have stagnated in this way, although the imperial city itself was fed largely with produce from Egypt, whose land was replenished annually with nutrient-rich mud from the Nile flood. Thanks to this wonderful source of fertilizer, the civilization of ancient Egypt had already endured for thousands of years.[†]

The deterioration in the quality of farmed soil has been reversed over the twentieth century in many regions of the globe, and population size has increased accordingly. Around seventy-five million tons of phosphate rock are mined each year and converted to fertilizer, an amount which probably exceeds the quantity released naturally from weathered rocks. While the use

---

* Biomass on land contains around two billion tons of phosphorus, far in excess of that in the sea, which is around 120 million tons.

[†] In China the problem was solved by growing rice, a crop that requires little phosphate and potassium, and by the careful return of all human dung to the paddyfields.

of phosphorus fertilizers no doubt increases the leaching from farmland to rivers, the net result has been a build-up of phosphorus in the soil.

Phosphate fertilizers

The discovery that phosphorus is the limiting factor to life on Earth was not without its controversial side. It took a while to unravel the role of phosphorus in soil enrichment and plant growth and then to determine which forms of phosphate were suitable for use as fertilizer.

It has been calculated that soil water contains only about 1 per cent of the phosphorus required to sustain normal plant growth for a season. However, this water is in contact with soil particles which contain phosphate and the laws of chemistry then come into play, so that as the phosphate is sucked up by plant roots more dissolves in the water to keep the concentration in the water constant. All kinds of soil particles can contribute to this process and in some cases it is not only chemical balance that maintains the supply but the action of microbes and enzymes that release phosphate from organic debris in the soil.

How much phosphorus do crops remove? Seed crops remove most because the seed stores phosphate, as inositol hexaphosphate, for when it germinates, so it is not surprising that a crop of beans will remove 11 kg of phosphorus from a hectare of land. Potatoes also remove a similar amount, while wheat takes 7 kg and sheep 1 kg and trees only 0.5 kg.

The eighteenth century had already seen major improvements in farming methods, including the introduction of four-field crop rotation which eliminated the need for land to lie fallow. In this method of farming, a cycle of crops could be grown on the same piece of land in a way that led to increased output. The sequence was: (1) human food, such as wheat, (2) animal fodder, such as turnips, (3) another human food, such as barley and (4) a crop such as clover, which could be eaten by animals or ploughed back in to act as fertilizer before the cycle began again. Much of the nutrients removed in crops was returned as manure and compost, but that which went to towns and cities was permanently lost from the land. Some artificial fertilizing of the land was carried out, such as the spreading of lime (calcium oxide), which was known to increase crop yields, for reasons we shall discover.

In theory, a particular phosphorus atom will move from soil to plant to animal and back to the soil in the form of excrement, and this cycle could continue for thousands of years. The underlying rock contributes to the pool of recycling phosphorus, but as water drains from the land to rivers, lakes and seas, a little dissolved phosphate is lost with it. In the oceans it might cycle through various forms of marine life, and there too it is replenished continually by run-off from the land. However, the sea also loses phosphate continually in a steady drizzle of detritus to the bottom, where it builds up in the sediment as insoluble calcium phosphate.

The result of all this activity is that there is a downward spiral of phosphate to regions where total blackness prevails and little life is to be found. Over most of the oceans the dilemma is that all the sunlight is at the top and all the nutrient phosphate at the bottom. Far from teeming with life, the oceans are sparsely populated. All is not lost, however, because there are some upwelling regions, where a little of this phosphate-laden lower water can return to the surface. Because of the way the Earth rotates and the ocean currents move, this generally happens on the west coasts of the continents. Despite this counter-current to the cycle, there is a net annual loss of around fifteen million tons of phosphate a year from the marine biosphere.

Useless fertilizer

In 1799, Erasmus Darwin, grandfather of Charles, wrote about plant nutrition in his book *The Philosophy of Agriculture and Gardening* and said that both nitrogen and phosphorus were essential components of plants and that these were absorbed through plant roots. He suggested that they should be added to the soil in the form of bone ash, manure and compost and recommended a search for phosphate rock deposits. In this, he was well ahead of his time, too far ahead to be taken seriously, and his advice was ignored.

In the early nineteenth century, scientists began to analyse the composition of plants and so discovered how important phosphorus was. Everything that lives contains phosphorus. Indeed, it became apparent that phosphorus might even be *the* element that governed fertility of the land. Bone meal began to be used as a fertilizer and this proved to be beneficial. 'Organic' farmers today still use bone meal, and this particular fertilizer remains popular with rose-growers. In 1833, the Duke of Richmond proved that the fertilizing value of bone meal resided not in the calcium it contained – in other words it was not another form of lime – but in its phosphorus component.

Its beneficial effects as a fertilizer could be speeded up by treating it with sulphuric acid, thereby converting the insoluble calcium phosphate to soluble calcium hydrogen phosphate and this much improved variety became known as 'superphosphate'.

In 1840, Justus von Liebig, the great German chemist who lived from 1803 to 1873, put forward the theory that the chemical processes that take place in soil were the basis of fertility and that, by influencing these, it should be possible to increase the productivity of the land. The British Association for the Advancement of Science commissioned Liebig to write a report on his ideas and this, published in 1840 under the title *Organic*

Chemistry and its Applications to Agriculture and Physiology, was presented at the association's Glasgow meeting that year. Scientific farming had begun and Liebig's message fell on fertile ground.

Liebig was writing at a time when the population theory of Thomas Malthus, the English economist, was current. Malthus foresaw that there must eventually be a catastrophe for the human race because of its capacity to reproduce 'geometrically', in other words we were multiplying, whereas food production could only increase 'arithmetically', in other words by the simple addition of more land devoted to farming. Clearly the one would eventually outpace the other and the human population might double, double and double again while the food to sustain such growth would become a commodity in limited supply. He theorized that eventually there must come a point where there were just too many mouths to feed and no more land to farm and the human race would be faced with mass starvation. Despite the improved farming methods introduced in the eighteenth century, the land could only yield a level determined by Nature. The way to avoid the coming crisis was to boost crop yields, although for most farmers it was still drought, plant diseases and insect pests which most affected productivity. Could fertilizing the land compensate for these?

Liebig thought it could. He knew that chemicals in the soil provided plants with their nutrients and reasoned that, by increasing their supply, we would increase the supply of food. Liebig was a master of chemical analysis and identified the minerals in soils and plants. He announced that plants needed carbon dioxide, water, ammonia, potassium nitrate, calcium oxide (lime), magnesium oxide, iron, phosphorus, sulphate and silicate. He also proposed his 'Law of the Minimum', which said that one, and only one, of these essential nutrients could be a limiting factor and it would be on the supply of this mineral that productivity depended. If it could be identified and the amount increased, then plant growth should go up accordingly. When enough of the nutrient had been added, the role of limiting factor would pass to another nutrient and it would then be necessary to increase that one . . . and so on.

Liebig suspected that phosphorus was *the* limiting factor. He also reasoned that animal manure could not in itself redress the lack of plant nutrients and that mineral-based fertilizer would be needed. He calculated the components of an ideal fertilizer and patented it. It was a failure – because it contained no nitrogen. Liebig deemed this element to be unnecessary in his mixture because he assumed, wrongly, that plants could get all the nitrogen they needed from the atmosphere. He was also wrong about phosphorus. He had included an *insoluble* form, calcium phosphate, in the belief that if he used a soluble version it would be washed from the soil by rain water. The first of these errors, about nitrogen, he remained oblivious to, but the second error he corrected in later editions of his book, realizing that its solubility was exactly what gave phosphorus the ability to boost plant growth.

Although Liebig was undoubtedly a great chemist, he was not an agronomist and his half-baked ideas about soil fertility brought him into bitter conflict with two Englishmen, John Lawes and Henry Gilbert (see box overleaf). Unlike Liebig they were *practical* chemists. The revolution that Lawes and Gilbert started was to lead to the development of modern methods of farming. These methods have doubled, tripled and even quadrupled the yields of crops, compared with those achieved by traditional farming methods.

Lawes had already started to manufacture superphosphate from bone meal in 1842 and suggested that this was a better way of applying phosphorus to the land than Liebig's way. Indeed it was. Superphosphate is composed of calcium hydrogen phosphate and calcium sulphate, but it sometimes has a lumpy texture which makes it difficult to spread. Even so, this new fertilizer

When great minds don't think alike

John Lawes believed that farming could be improved by the application of science and to prove his point he set up an experimental research station at Rothamsted, 40 kilometres (25 miles) north of London, in 1843, and with the help of Henry Gilbert he began to experiment with fertilizers. Carefully measured plots were planted with crops, fertilized in various ways, harvested and the yields recorded. Lawes knew that the soil of Rothamsted was deficient in nutrients as a result of centuries of inefficient agriculture. Gilbert was a chemist who had trained under Liebig at Giessen, in Germany, and he was keen to apply the methods for improving soil as advocated by Liebig and especially to test his patented fertilizer. Lawes and Gilbert quickly discovered that adding Liebig's patent fertilizer to the land did not work. They knew the amount they were using was enough to provide the phosphorus and potassium required for the expected larger crop and wondered if, perhaps, the low yield was still somehow due to the impoverished level of these nutrients. However, adding yet more fertilizer still did not increase yields accordingly. Could the great Liebig have got it wrong? They repeated their field trials yet again and still the promised bounty did not manifest itself in improved crop yields.

In fact they found a lot that Liebig said about soil fertility was wrong, and began publicly to say so, thereby starting a slanging match that went on for twenty years, and which eventually descended to hurling insults. Liebig was furious that his ideas were being challenged by a mere farmer and saw Gilbert's involvement with Lawes as a kind of betrayal. At one point an incensed Liebig referred to them both as 'vile vermin'. But such was the stature of Liebig in the scientific world that his theory continued to be believed, despite the evidence that

Rothamsted was accumulating. Thankfully Lawes and Gilbert were not cowed by the scorn which Liebig poured on their efforts and in the end the glory was to be theirs. They were proved right: there was something wrong with Liebig's fertilizer. It was not addressing the second most important limiting factor after phosphorus, which was nitrogen.

boosted crop yields and, by 1850, it was also being made by treating phosphate rock with sulphuric acid.

An even better variety, called 'triple superphosphate', became available around this time and was made by dissolving bone meal in phosphoric acid. Triple superphosphate can provide more phosphorus on a weight-for-weight basis than superphosphate because it consists solely of soluble calcium hydrogen phosphate, and it is easier to handle and apply to the land because it is a free-flowing granular material.

On the original plots of land at Rothamsted the annual applications of fertilizer recommended by Lawes and Gilbert are still carried out to this day, in what has become the longest-running agricultural experiment of all time. For over 150 years, wheat, grass and other crops have been grown on the same soil, with exactly the same amounts of different fertilizers added. There is even a plot of land that has had no fertilizer applied to it at all and this manages to produce a small crop of one ton of wheat per hectare (two and a half acres), which is about the medieval yield.*

The plot fertilized with inorganic fertilizers, which supply nitrogen, phosphorus and potassium, produces six times this

---

* Whereas peasants fertilized their land with manure, this crop now gets its nitrogen from the nitrogen oxides dissolved in present-day rain water. These come mainly from car exhausts.

amount and even farmyard manure liberally spread will boost the yield to five times the non-fertilized amount. The trouble with animal manure, as with all traditional farming methods, is that other land has to be used to grow fodder for the animals that produce the manure. And this is the dilemma at the heart of all traditional ('organic') farming: some land somewhere must lose nutrients.

When the ton of wheat from the unfertilized plot is harvested, it removes about 5.5 kilograms of phosphorus. So how is this replaced? It is constantly being removed year by year, but replenishment must come from somewhere. In fact, it comes from the breakdown of the minerals that form the soil. As rock weathers under the action of rain, which is slightly acidic, and frost, it breaks down and releases its minerals. A hectare of soil will hold enough phosphorus and release enough of it to sustain a modest crop yield, as the experiments at Rothamsted have shown.

In fertile regions of the world, the element that can best boost crop yields is nitrogen, which is why intensive agriculture in advanced countries is so successful. The nitrogen in the air is useless as a fertilizer and has to be converted to compounds such as ammonium nitrate before it can be absorbed by plants. Some micro-organisms can 'fix' nitrogen directly from the air, some convert it to ammonia, while others are able to turn it into nitrate, which plant roots can take up and use to make amino acids and proteins. These micro-organisms are responsible for supplying this essential nutrient to the world's land-based food chains, but there is a limit to the amount they can fix. Algae do a similar fixing of nitrogen for organisms that live in the sea.

We know from Liebig's Law of the Minimum that spreading excess nitrogen fertilizer on the land would be of little use unless we had first removed the limiting factor of phosphorus. In over-cultivated and exhausted soils, we need to add excess phosphate for many years until maximum fertility is restored. In Australia, such has been the depletion of phosphorus from the land over aeons that in some field tests it was necessary to add almost half a ton of phosphate fertilizer to a hectare of land before phosphorus ceased to be the limiting factor to plant growth.

Reaping the rewards of fertilizer phosphate

The phosphate in any ecosystem is the key to maximizing crop yields. Too little phosphate is likely to be the problem in developing countries, too much is the problem in developed countries. The former can be solved by helping farmers to buy phosphate fertilizer, the latter by carefully monitoring the soil and maintaining soil phosphate just above the critical level for optimal production. This allows for the steady drain of phosphate from the land to streams and rivers.

Once humans realized the importance of phosphorus to plants, something could be done about it. Farmers of the nineteenth century scattered superphosphate and triple superphosphate and reaped the benefits accordingly. Farmers of today can avail themselves of even better phosphates, especially ammonium and potassium phosphates, which add the other two nutrients that need endlessly to be replenished in any soil. They can still avail themselves of some of the over-abundance of fertilizer phosphate applied many years ago, although this excess was applied in the belief that not all phosphate was available to crops.

Analysis of the phosphate in the soil suggested that relatively little of the surplus applied one year would be available for the following year's crop. So where was it hiding? Plants need phosphate that is soluble, and the amount of this present in the soil can be judged by taking a sample, shaking it with water, filtering off the solid particles and analysing the solution for the phosphate it contains. When this quantity is considerably less than the amount that has been added, it must either mean that rain water has carried the rest away, or that it has somehow become insoluble and is lurking in the filtered-off particles of soil. In fact, it is due to a bit of both, but typically most has been rendered insoluble.

Soil analysis could detect phosphorus in the soil particles but, being insoluble, it was assumed to be unavailable to plant roots. The soluble phosphate of the fertilizer has been rendered inactive by reacting with metals present in the soil, in particular calcium, iron and aluminium, which form insoluble phosphates. For most of the twentieth century, farmers were advised by government departments of agriculture to add more than double the amount of phosphate required by a crop. However, insoluble phosphates are not as unavailable to plants as the men at the ministry had assumed, and we now know that this pool of phosphorus is capable of being tapped by plant roots.

Seeds need phosphorus when they start to grow, which is why they contain a store of phosphorus for this purpose. The plant creates its store by making inositol hexaphosphate, a molecule with six phosphates. As the seed starts to grow, it produces enzymes that can unlock this store, thereby releasing the phosphorus to the growing roots. Eventually these develop the ability to extract phosphate from the soil, where all kinds of phosphorus compounds are present. Most is present not as soluble inorganic phosphate but as plant debris and detritus of organisms that live in the soil. Plant roots are efficient at extracting all varieties of phosphorus, even to the extent of rendering the surrounding soil slightly acidic to facilitate solubilization and producing phosphatase enzymes in order to release phosphorus from organic debris. They are aided by bacteria in the soil, which are also able to solubilize phosphorus.*

---

* Another way to solubilize phosphates is to spread lime on the land. Aluminium in soil has a particular affinity for phosphorus, especially if the soil is slightly acid. Liming soil has the effect of releasing the hold that

During the Second World War, the farmers of Denmark, who were unable to import phosphorus fertilizers, found to their surprise that they were able to reap crop yields at pre-war levels and overall food production even rose slightly during the five years they were occupied by the Nazis. They were in fact drawing upon the phosphate that had been built up over the previous decades, and while theory suggested that fertilizer phosphate was no use once it had been converted to insoluble phosphate, plant roots knew otherwise.

Phosphorus's planetary imprint

Animals need phosphorus to form the bones of their skeleton, which mainly comprise calcium phosphate. Most phosphorus that adult animals take in they excrete back to the land, to go round the cycle again. Even their skeleton will eventually disintegrate under the action of weathering and return its store of phosphate to the common pool, although this may take a very long time and sometimes may never happen – which is why living things can leave an imprint on the planet lasting hundreds of millions of year. If a dead animal falls into a swamp, or its body is engulfed by mud, its skeleton may be preserved for ever, even turning up a hundred million years or so later, as we see with dinosaur bones. Bone survives in rocks for millions of years because it is mainly calcium phosphate, a mineral that is tough and insoluble.

What then preserves fossilized leaves and other organisms that have no calcium phosphate skeleton? The answer, again, is calcium phosphate. How this happens was explained in 1994 by a team of scientists at Bristol University headed by Dr Derek

---

aluminium has. Iron too has the ability to attract phosphorus, but in this case the bond between them is too strong to be broken easily by calcium.

Briggs. For mineralization to occur oxygen must be excluded, so the ideal conditions can be found in the muddy sediment at the bottom of a stagnant pool. The Bristol team were able to show that when a leaf or dead organism finds itself in this muddy sediment it takes only a few weeks for it to become fossilized. Bacteria in the mud invade the tissue and use the calcium and phosphate from the dead organism's cells to construct an outline of microscopic particles of calcium phosphate. If the mud remains undisturbed it will solidify and eventually become rock, at which point a perfect fossil record will be preserved for ever.

The end of the world

The abundance of life on planet Earth is regulated by many factors, but the essential element most likely to be in short supply on heavily farmed land was phosphorus. As we have seen in this chapter, phosphorus moves inexorably through a downward spiral which drains it to the bottom of the deepest oceans. This natural trend of locking phosphorus away has been reversed by humans using artificial fertilizers. At the start of this chapter, there was a quotation from that great popularizer of science, Isaac Asimov. He concluded his chapter on phosphorus with the words:

> We may be able to substitute nuclear power for coal, and plastics for wood, and yeast for meat, and friendliness for isolation – but for phosphorus there is neither substitute nor replacement.

That was in 1975, when the problems of the world seemed very different from those we face today. Then people were worried about future energy supplies, fast-depleting natural resources, lack of food to feed a rapidly expanding population and a supposedly imminent New Ice Age. Those concerns have given way to others which are centred on pollution, disease, wildlife and biodiversity. However, what Asimov said about phosphorus still holds true.

Left to itself, the Earth would gradually have deposited more and more of life's limiting factor at the bottom of the deepest oceans. The abundance of living things on land and in the upper sunlit regions of the oceans would have dwindled accordingly. The world would have declined into a semi-barren old age with thin vegetation sustaining only a few creatures. The inbuilt limitation of phosphorus has been checked – at least for the time being.

13. Oh, shit!

Tide gets clothes cleaner than any soap!
Any soap?
Yes, *any* soap! Tide gets clothes cleaner than any soap!
T-I-D-E, Tide!

Excited voices trilled this ditty on the radios of the USA in 1946 when the first phosphate-containing detergent, Tide, was launched. The world was about to change and seemingly for the better, thanks to phosphorus. Having shown what benefits it could bring to agriculture, it was now about to deliver some real benefits in the home.

The story begins in Cincinnati, Ohio, where in 1833 an English candle-maker, William Procter, and an Irish soap-maker, James Gamble, married two sisters. In 1837 the brothers-in-law went into a partnership as the Procter & Gamble Company, and so began a business that was to revolutionize the way we live, or at least the way in which we clean our homes and clothes. The company's business was principally the production of soap for the US market, but it slowly expanded into other countries and other markets, notably healthcare (Vicks) and beauty products (Max Factor and Old Spice). In the 1930s Procter & Gamble even went into the business of making dramas for radio stations, within which they would advertise their wares, hence the term 'soap opera'. Today they still produce soap operas for television, in New York.

Until the twentieth century there was no reason to believe

that phosphorus would have a role to play in household cleaning, but then sodium tripolyphosphate (STPP) appeared on the scene. This boosted the power of soap powders, not by making the soap any better but by making the wash water softer. It was so successful that within ten years of its first use, in 1946, all detergents contained it. However, a further ten years down the line and STPP was cast as the world's number-one pollutant. What had gone wrong?

In fact, it had appeared to remove the brake that phosphorus applies as the limiting factor in the aquatic environment. The abundance of living aquatic-based organisms that resulted – the algae – was unexpected and unwanted and the ecosystem itself could not cope with such swift changes. Lakes were awash with foul-smelling green slime. The answer appeared to be simple: ban phosphates in detergents, said environmentalists; but scientists knew this could only be part of the answer.

Phosphorus in the form of phosphate enters our homes in many of the products we purchase each week at the supermarket: meat, desserts, drinks, toothpaste, detergents and dishwasher tablets. These phosphates exit down the kitchen waste pipe or are flushed away down the lavatory pan and end up at the local sewage-treatment plant – and that is where the trouble began. The human body excretes around 0.8 kg of phosphorus a year and detergents used to add about three times as much per person, so that a household of four people could generate about 13 kg of waste phosphorus a year.*†

In this chapter we will look at the various forms in which phosphate comes into our lives and the problems it causes when the waste from our homes enters the environment.

---

* Equivalent to 40 kg of phosphate, enough to fertilize a 5-hectare crop of wheat.

† Current usage of phosphates has reduced this to 5 kg.

Phosphorus in food

The average person has in their body about 3.5 kg of calcium phosphate, some of which is taken from the mother while in the womb and the rest is accumulated from food. Although it is present in all living things, by the time some foods are processed they contain little phosphorus, and a few such as sugar contain none at all. In theory, it would be possible to design a diet rich in carbohydrates, fats and protein but which excludes phosphorus entirely. Needless to say, we could not raise children on it, nor even last very long on it ourselves. On the other hand we don't need to make elaborate plans to ensure we get enough phosphorus, even when we are young and growing rapidly, because so many foods contain it. Even a junk-food meal of hamburger, fries and cola contains more than enough phosphorus for our body's daily needs. Phosphoric acid is added to drinks like Coca-Cola to give them a pleasant tang, and unlike other acids, such as the citric acid which was originally put in colas, it does not interfere with the taste of the flavour ingredients.

The dietary reference value for phosphorus, the term used to describe the amount we should take in each day, is equivalent to 800 mg of phosphate for an adult, roughly the same as that for calcium, a mineral of which we are far more aware and which we know we need to keep at a high level in our diet. The phosphorus in our body is mainly found in conjunction with calcium, and 80 per cent of the total body content is contained in the skeleton, with the remainder distributed in the soft tissues. Blood levels in adults vary between 2.0 and 4.3 mg of phosphorus per 100 ml, while children have much higher levels of between 4.0 and 7.0 mg per 100 ml, which indicates how important this element is for their growth and development.

Phosphorus plays a key part in almost all the essential functions of a cell, such as growth, structure, metabolism and

---

Phosphorus in foods

The amount of phosphorus that is naturally present in food varies considerably, but it can be as high as 370 mg per 100 g in liver, or can be very low, as it is in vegetable oils. Lean meat generally contains 180 mg per 100 g, wholemeal flour has 340 mg whereas white flour has 130 mg. However, the phosphorus that is in the bran component of wholemeal flour is inositol hexaphosphate, which we cannot digest; the same is true of nuts, which also contain a lot of phosphorus, but three-quarters of this is inaccessible. Eggs are also a good source of phosphorus with 220 mg per 100 g – all of it digestible – as is that of cheese, which can be one of the best with 500 mg. This is partly because milk has a high level and partly because phosphates are added to certain types of processed cheese.

---

reproduction. Our genes and reproductive capability stem from the phosphorus-based polymer DNA. A cell needs phospholipids, such as lecithin, to form its outer membrane, and adenosine triphosphate (ATP) in order to enable it to function. ATP is the molecule that drives the myriad chemical reactions which form most of the cell's components, as well as being the molecule that allows the cell to derive all the energy it requires from glucose. Outside the cells are other phosphorus molecules such as cyclic-AMP,* which transmits information around the body, and triggers the release of hormones.

Bone phosphate represents an enormous store of phosphorus. We think of bone as inert but in living things it is constantly degraded and restored at millions of remodelling sites throughout the skeleton by cells called osteoclasts and osteoblasts. In this

---

* Short for cyclic-adenosine monophosphate.

way bone carries out its secondary function of maintaining a steady level of calcium in the blood. The skeleton also represents a reservoir of phosphorus that the body can tap into at any time, so there is never a shortage. In the nineteenth century nutritionists thought differently, which is understandable because at the time there was a high incidence of bone deformities such as rickets. Consequently various foods were 'fortified' with phosphates in the belief that a lack of this nutrient was responsible. In fact, rickets was due not to a lack of phosphate, or calcium, in the diet but to a deficiency of vitamin D. While humans do not lack phosphate, other animals can and this is why dicalcium phosphate is added to cattle feed as a mineral supplement.

Since phosphate is an essential part of our diet, it is perhaps not surprising to find that phosphates are considered safe when used as food additives. Even when the food additive is STPP, which is the commonly used dishwasher and laundry detergent component, it is deemed safe because the body is plentifully equipped with phosphatase enzymes that can break down this complex phosphate into simple phosphate. Additives are used to improve food texture and regulate acidity, but their use has not been without controversy.

'Why sell meat when you can sell water?' Perhaps the benefit of using phosphates was never explicitly stated as such, but those in the business of selling chicken, bacon and ham quickly cottoned on to its potential. A little polyphosphate in meat makes it possible for a lot more water to be retained, in some cases double the natural quantity. However, the reason polyphosphates are added to such meats is to improve their shelf-life and to make them easier to slice. The first of these benefits comes about by tying up the iron and copper and so denying these essential elements to spoilage bacteria as well as preventing them catalysing the attack of oxygen from the air. The second of these benefits is achieved by enhancing the water-holding ability of the muscle protein. Consumers actually prefer meat to be moist and tender, in other words more succulent, and this is what polyphosphates will deliver.*

Ham is prepared with disodium phosphate and STPP, which are added to the brine solution which cures the ham. Originally, the phosphate was there to regulate the acidity of the pickling solution, but it was found that the resulting cured ham, especially if it was also smoked, was juicier and more tender. Moreover, unlike ordinary smoked ham, which loses 10 per cent of its weight during the process, the phosphate-treated ham lost no weight at all.

The permitted phosphate additives used most frequently in foods are given in Table 13.1. Some are salts of phosphoric acid, H_3PO_4, some of diphosphoric acid, $H_4P_2O_7$, some of triphosphoric acid, $H_5P_3O_{10}$, and some of polyphosphoric acid, $H_nP_nO_{3n}$. These acids have multiple acidic hydrogens (H), most of which are neutralized by sodium, potassium or calcium before they are used as food additives.

In some cases, the additive is used because of its residual acid content as well as its phosphate content. For example, disodium diphosphate, $Na_2H_2P_2O_7$, is used as a raising agent for doughnuts, biscuits and baking powder because its acidity (H) reacts with carbonate to form bubbles of carbon dioxide. It is also added to peeled potatoes and canned fish where it is the diphosphate that is the active part; this sequesters metals like iron, calcium and magnesium which can cause discoloration and crystal formation.

The use of phosphate food additives goes back to the mid-nineteenth century when they were first introduced as raising, or leavening, agents, to give baked foods, such as cakes, pastries and biscuits, a lighter texture. The first baking powder, introduced in the 1850s, was a mixture of potassium hydrogen tartrate (also

---

* The fact that frying bacon cured with polyphosphates causes it to shrivel up before your eyes is surely a small price to pay.

Table 13.1: Phosphates used as food additives

Food additive	Used in
Phosphoric acid	colas, cooked meats, sausages, ham, cheese
Sodium phosphate	cheesecake mix, lemon-pie filling
Disodium phosphate	spreads, cooked meats, sausages, instant desserts, cheese slices
Trisodium phosphate	cooked meats, ham, sausages, cheese spread
Potassium phosphate	cup soups, trifle mix, dessert topping
Dipotassium phosphate	non-dairy powdered coffee creamers
Calcium phosphate	cheeses and crisps
Ionsine 5'-disodium phosphate	rice-based snacks and gravy granules
Sodium aluminium phosphate	packet cake mixes
Calcium diphosphate*	short pastry mix and baking powder
Tricalcium diphosphate*	cake mixes
Sodium diphosphate*	bread, catering-grade whipping cream, cheese, condensed milk, dried milk products
Sodium tripolyphosphate	cheeses, packet soups, condensed milk
Potassium tripolyphosphate	canned hot-dog sausages
Sodium/potassium polyphosphates	poultry, ham, bacon, canned custard, reduced-sugar jam, cheese, frozen fish fingers
Calcium polyphosphates	cheese
Ammonium polyphosphates	cheese

* Diphosphate is also known as pyrophosphate.

known as cream of tartar) and sodium hydrogen carbonate (also known as bicarbonate of soda). When baking powder was added to the flour and it was then moistened and heated, it would produce lots of tiny bubbles of carbon dioxide gas.

A better baking powder was devised by a professor of chemistry at Harvard University, E. N. Horsford, who put together a mixture of calcium dihydrogen phosphate as the acid and sodium hydrogen carbonate. In 1873, self-raising flour first appeared on the market with Horsford's mixture already blended in, but this was not popular because the quality of the calcium hydrogen phosphate was unpredictable. It was only in the 1930s that quality control ensured a reliable product, after which self-raising flour became the preferred flour for home baking.

Phosphates are essential for producing processed cheese, and the main ones used for this purpose are disodium phosphate or sodium aluminium phosphate. Processed cheese is mainly Cheddar cheese that has been pasteurized at 70°C to deactivate the fermentation bacteria. To prevent the butterfat from separating, it has to be emulsified, and the best way of doing this is to add around 2 per cent of disodium phosphate, which is readily soluble in the melted cheese. Phosphates are effective emulsifying agents because they make proteins more soluble, especially the milk protein, casein. The result is an homogeneous melt that is extruded as a ribbon and cut into slices. Millions of these slices are eaten every day in sandwiches or cheeseburgers.

Disodium phosphate is also added to evaporated milk to emulsify the fat, and it also gives the necessary calcium-to-phosphate balance that prevents this type of milk from becoming semi-solid when it is stored. Tetrasodium dihydrogen phosphate is another phosphate that is used to disperse solids in milk and is the secret behind thick milkshakes and 'instant' desserts such as Angel Delight.

Canned seafoods, such as lobster, crab, salmon and tuna, have a tendency to form crystals of struvite, which is magnesium

---

Phosphoric acid

Phosphates used in food need to be extremely pure and are made from pure phosphoric acid. There are two ways in which this is made by the chemical industry. 'Thermal acid' is the phosphoric acid produced by burning elemental phosphorus to form phosphorus oxide, P_2O_5, and then immediately reacting this with water to form the acid, H_3PO_4. The process generates a considerable amount of heat, hence the name 'thermal acid'. This distinguishes it from the acid made by dissolving phosphate rock in sulphuric acid, which is called 'wet acid'. At one time this could not be used in processed foods because it was contaminated with the impurities present in the rock. Wet acid could be purified to a certain extent – and today it can be rendered pure – but for a long time it was considered only suitable for the manufacture of fertilizers and for industrial use.

---

ammonium phosphate ($MgNH_4PO_4$) over a period of time and these look like tiny slivers of glass. The formation of these can be prevented by adding sodium hydrogen diphosphate or sodium polyphosphate, which will sequester the magnesium and render it unavailable for crystal formation.

Detergents

Trisodium phosphate (TSP) was for many years the miracle ingredient in household cleaners because, when dissolved in water and used to wipe surfaces, it would react chemically with any fat and grease it came into contact with. It converted them

---

Hard and soft water

Rain water is very soft, but most water used in the home comes from rivers, lakes and underground supplies. This water may have dissolved minerals from the soil and underlying rocks, especially calcium and magnesium, and as such is then described as hard, the degree of hardness depending on the amount of dissolved minerals. Hard water reacts chemically with soap molecules to form insoluble soap scum or soap curd, which reduces the effectiveness of soap and leaves deposits on both the clothes being washed and the container they are washed in. However, hard water does not form an insoluble scum with detergent molecules, which is why these are much better for washing clothes.

---

into water-soluble glycerine and soap.* Today TSP is considered too risky to be used in general cleaning products, but it is still used to remove food-poisoning germs from raw chicken. Bacteria such as salmonella, which cling to the surface layer of fat, are washed away when a carcass is sprayed with a solution of TSP. Chemists at the Rhône-Poulenc labs in Cranbury, New Jersey, patented the process, which received approval in 1992 from the Food and Drug Administration (FDA).

Soap is the sodium salt of a fatty acid, for example sodium stearate or sodium palmitate. These chemicals are fine for removing dirt from clothes, provided the water is soft (see box).

Until the Second World War, soap produced from animal fats or vegetable oils was used for most cleaning around the home. Today, we prefer the more efficient, if synthetic, surfac-

---

* Unfortunately, it did the same to the natural oils of the skin, leaving the hands chapped and sore.

tants, which are made from oil, although this oil may come from natural sources as well as from petroleum. Both soap and surfactants attach themselves to dirt and grease, solubilizing these in the water so that they can be washed away. But both are vulnerable to calcium and magnesium in the wash water, visibly so in the case of soap. By softening the wash water, in other words inactivating the calcium and magnesium, a little soap or surfactant can be made to go a long way.

At first sight, it would seem unlikely that phosphate could be of any use. After all, calcium phosphate is even more insoluble than the calcium salts that constitute soap scum, that is calcium stearate and calcium palmitate. This is a problem that Nature has also had to wrestle with. Both calcium and phosphate are needed by living cells and both must be kept in solution. The answer is to join two or more phosphates together, making them impervious to calcium; these multiple-linked phosphates actively cluster around calcium and keep it afloat. With the benefit of hindsight, we can see now why adding linked (poly) phosphates to detergents would improve their washing abilities.*

The first compound to be used for this purpose was tetrasodium diphosphate, $Na_4P_2O_7$, with two linked phosphates. This material, made by heating disodium phosphate at 500°C, was a fine white powder that, when added to detergents, boosted their cleaning ability. The diphosphate acts as a 'builder', a term used to describe ingredients that enhance the action of the soap or surfactant. Before long, though, an even better builder was discovered: sodium tripolyphosphate (STPP), in which three phosphates are joined together, just as they are in the naturally occurring molecule ATP.

The ability of STPP to bind to metals was already known about in 1930 when Procter & Gamble first became interested in

---

* Modern detergents contain lots of other components, such as enzymes, bleaches, brightness enhancers, perfume and so on.

it, but nothing was developed until 1938, when David Byerly joined the company. He worked on a project to find builders for soap powders. It was while he was researching detergent formulations for hotel dishwashers that he tried adding STPP and noticed that the dishes came out far cleaner. Byerly became fascinated with STPP, so much so that even when he was ordered to stop working on it because the project seemed to be getting nowhere he continued to research it in his own time. He tried adding it to laundry formulations and the results were dramatic. But by then the US was at war and the project was shelved until hostilities ended.

After the war Procter & Gamble invested heavily in the new 'builder' and in 1946 they launched Tide. They advertised it as 'the new washing miracle' which produced 'oceans of suds' capable not only of washing clothes whiter but of leaving dishes sparkling clean. By 1950, Tide was outselling all other brands.

As a builder, STPP does several things: it softens water by sequestering calcium and magnesium; it keeps the wash water slightly alkaline, at an optimum pH of between 9 and 10 which gives good cleaning results without damaging the skin. STPP keeps dirt in suspension once this has been washed off clothes and it boosts the action of the surfactant so that less is required. On the other hand, it also greatly increased the amount of phosphate ending up in the sewers, sewage works, rivers and lakes, and was the obvious target to blame for causing water pollution because it constituted a third of the weight of most popular brands of detergent.

In the 1970s phosphate eventually became *the* pollutant and all products that contained it were called into question – even toothpaste. This was once heavily dependent on calcium phosphate, which acted as an abrasive. (It has now been partly replaced by other abrasives, such as silica.) Many toothpastes also contain a little phosphate in the form of monofluorophosphate, which provides the fluoride that strengthens teeth and

---

The strange attraction of fluoride and phosphate

When Domenico Pini Morichini analysed the enamel of fossilized elephants' teeth in 1802, he was surprised to find that although it consisted mainly of calcium phosphate, it also contained a lot of calcium fluoride. This was the first evidence of a link between these two minerals, although in the case of fossilized bones the fluoride accumulates over aeons. The most insoluble form of calcium phosphate is hydroxyapatite, which has the chemical formula $Ca_2(PO_4)_3(OH)$. When this comes into contact with fluoride, the hydroxy part, the (OH), can be displaced by fluoride, which binds more strongly and moreover makes the mineral even more insoluble as fluoroapatite, $Ca_2(PO_4)_3(F)$. This is why fluoride is added to water supplies and why we should brush our teeth with fluoride toothpaste. Tooth enamel is pure hydroxyapatite, which may be very hard but is slightly susceptible to attack by strong acids, such as those produced by bacteria acting on sugar. Exchange the OH for F and you create a tooth enamel that is even stronger and more acid-resistant, which explains why those who grow up in areas where water is fluoridated invariably have better teeth.

---

fights tooth decay. Fluoride is attracted to the calcium phosphate which makes up tooth enamel (see box). This affinity explains why most of the world's phosphate minerals incorporate a lot of fluoride and are found as fluoroapatite.

Eutrophication

The word eutrophication is derived from the Greek and means 'well nourished', but it really means 'too well nourished'; it applies to an aquatic system which is over-supplied with nutrients. In the 1950s problems developed in certain lakes and inland seas, especially the Great Lakes of North America, which were suffering from eutrophication in the form of a perpetual algal bloom. Eutrophication of this kind was a disaster: the living algae were blocking out the light of the sun, while dead algae drifted to the bottom where their decay was using up the oxygen dissolved in the lake's water, which is essential for other creatures, especially fish. Lakes in Europe and the US became green and smelly, devoid of fish, unfit as sources of drinking water and unimaginable as places of recreation.

The finger of suspicion pointed at phosphate, for reasons we saw in the previous chapter, since this element was known to be *the* limiting factor for life in the aquatic environment. At a series of hearings into the problem held in Washington DC in 1969, Congressman Henry Reuss summed up the causes of eutrophication and offered a quick way of solving the problem: remove the STPP from detergents. The eutrophication of the Great Lakes seemed to have resulted from the increasing use of the new STPP-based detergents. The decision to ban phosphates was an easy form of action. It appeared that little could be done about the other contributors to the increased level of phosphorus in the Great Lakes: the phosphate draining from farmland and that contained in human sewage.

Research into aquatic systems demonstrated the effect that phosphate could have. When radioactively labelled phosphorus* was added to a typical lake, the results were amazing. Within a minute, half of the phosphorus had been absorbed by organisms such as algae and within five minutes the entire amount had been used. Algae are able to take in excess phosphorus and can store enough to see them through three generations.

---

* Radioactive phosphorus isotope 32 allows its course to be followed by monitoring the radiation emitted by organisms which absorb it.

Algae bloom naturally in lakes in spring and summer, but such blooms usually last only a few days and quickly fade. The spring bloom occurs when there is enough warmth and sunlight for the algae to multiply, drawing on the phosphorus that has built up in the winter. During the summer months, the phosphorus gradually sinks to the deeper, colder layer at the bottom of the lake and builds up there. Then the autumn bloom occurs when the surface water of the lake begins to cool, which allows the lower layer to rise, bringing its phosphorus to the top, and the algae feast again. However, given a continuous supply of phosphate, algae will bloom all summer long.

In June 1977 the US Environmental Protection Agency published a sixty-two-page paper entitled 'Detergent Phosphate Ban' which advocated an end to the use of phosphates in detergents in the Great Lakes Basin. In 1971 the Agency had advocated a different form of phosphorus control, which relied on chemical treatment of municipal sewage and industrial wastes, but the policy had clearly failed to achieve the desired result. Lakes Erie, Michigan and Ontario were still badly eutropic, with increasingly frequent algal blooms. Masses of rotting vegetation were piling up on the beaches, while in the lakes huge fish kills were taking place due to oxygen depletion. Some scientists warned that irreversible damage was being caused and that the Great Lakes might never be restored to their former glory.

In 1972, the US and Canada agreed to a limit of 1 mg of phosphorus per litre in all water being discharged into the lakes, but algae can still bloom even with such low concentrations. More needed to be done. There was talk of lowering the level to 0.1 mg per litre, and it was proposed that detergents containing more than 0.5 per cent phosphorus by weight should not be sold. In support of such a course of action, the report's authors could point to what had been achieved by the State of Indiana, which banned phosphates in detergents in 1971. This had resulted in reduced phosphate levels for both inland lakes and rivers. Inevitably, a detergent ban on phosphate came into force in many states of the USA and in provinces of Canada in an attempt to revive the Great Lakes.

In other parts of the world where polluted water was found, blame also fell on phosphate, so much so that by the 1980s this element was generally regarded as little more than an environmental pollutant. Phosphate was considered to pose such a danger to natural aquatic systems that all new sewage plants were equipped with phosphate-removal facilities and older plants were adapted for this function too.

Bans and restrictions on phosphate in detergents were imposed in the Netherlands, Norway and Germany. Only in Sweden were phosphates actively encouraged. When Germany introduced phosphate restrictions, the consumption of phosphate for domestic use fell from 70,000 tons per year in 1970 to 20,000 by 1988. In France, the level of phosphate in washing powders was cut from 30 per cent in 1990 to 25 per cent in 1991 and 20 per cent in 1992. In Britain, most products contained 30 per cent, and dishwasher formulations contained 40 per cent, but, as there was little eutrophication of its rivers and lakes, the Government took no action to limit phosphates, beyond calling for more research to be done.

Nevertheless, eco-friendly 'phosphate-free' detergents appeared on British supermarket shelves as they did in many countries, but they were never able to capture more than a small fraction of sales. Most who tried them soon returned to conventional washing powders, observing that far more of the eco-friendly product was needed to achieve the same cleaning performance. In any case, the established brands had been reformulated into concentrated varieties to give cleaner washes using less powder at much lower temperatures and with zeolite water-softeners replacing STPP. Yet, while certain brands of detergents disappeared from supermarket shelves, phosphate-containing dishwasher powders remained because no other chemical would do the job as well.

Phosphate, like elemental phosphorus before it, seemed a devil in disguise and the less we had to do with it, the better. But redemption was at hand.

Phosphate vindicated

In January 1994 *The Phosphate Report* was published by Land-bank Environmental Research & Consulting of London. This was a cradle-to-grave study of both phosphate and zeolite by a group of independent scientists drawn from British universities and European institutions. Zeolites were the detergent builders replacing phosphate and they were assumed to be far more environmentally friendly. The report contradicted this belief; phosphate was not the eco-villain it was reputed to be. Some refused to accept the report's findings because the investigation had been sponsored by both a major phosphate producer, Albright & Wilson, and an industry-based organization known as the Centre Européen d'Etudes des Polyphosphates. However, great care had been taken to ensure that the report was unbiased, and in fact one of its authors, Dr Bryn Jones, was a former director of Greenpeace.

The report took into account all environmental costs including mining the raw materials, industrial production, energy input, transport costs, use by consumers and environmental pollution. The conclusion stated that there was little to choose between the polluting phosphates and the replacement zeolites in terms of their environmental impact. For the anti-phosphate campaigners, this came as a blow, but they felt they could ignore the report as it had been industry-sponsored.

Then, in May 1995, *The Swedish Phosphate Report* appeared – again produced by the Landbank Environmental Research &

Consulting Agency but with a larger panel of experts. It actually went further and claimed phosphate builders were as 'green' as other builders and explicitly recommended that those European countries that had banned or restricted the use of phosphate should now consider repealing such bans and restrictions. According to this report, phosphate was the best environmental option for water-softening systems and scored better than the alternatives in all circumstances of detergent use. The alternatives included not only zeolites, but another phosphate-free detergent builder, the polycarboxylic acids. The Swedish report concluded that the best way to protect rivers and lakes from eutrophication was to improve waste-water treatment. In 1997, the Nordic Swan eco-labelling board gave its seal of approval to STPP as an environmentally friendly component of detergents. For dishwasher detergents the amount of STPP can be as high as 53 per cent and still gain the Nordic Swan eco-label.

It was all rather unsettling for the environmentalist lobby. For thirty years, phosphate had been denigrated by ecologists and politicians as a major source of aquatic pollution, and both groups had encouraged alternative detergent builders and discouraged the use of all phosphates. Obviously, it was not possible to return to the years of profligacy in phosphate use, but partial rehabilitation was now deemed possible, provided it went hand in hand with the recovery of phosphate from waste waters. But could this second industry-funded report be trusted? Indeed it could, and support for it came from an unexpected independent quarter.

In the mid-1990s investigation into the algal pollution of lakes was being carried out by the prestigious Netherlands Organization for Applied Scientific Research, in conjunction with the University of Savoy in France and the University of Alicante in Spain. Martin Scholten led the seven-year investigation which came to the same conclusion as *The Swedish Phosphate Report*: that removal of phosphate from detergents was pointless and ill-advised. Water pollution was a far more complex problem than early environmentalist activists had assumed and it was proven to depend on a variety of factors, among which phosphate was perhaps one of the least important. The main cause was demonstrated to be industrial pollution by heavy metals, oils and insecticides, as all of these substances kill the zooplankton that feed on the algae.

The critical factor in eutrophication is the lack of zooplankton – tiny organisms, such as water fleas – which eat the algae. It was proven, by adding phosphate to unpolluted lakes, that phosphate is not solely responsible for eutrophication. An upsurge in algae did result but this encouraged the zooplankton to multiply and these in turn provided more food for fish. The key to maintaining a natural cycle of checks and balances is to have healthy zooplankton. When these are affected by pollution from industry and farming, they cannot cope with a sudden increase in algae and so the bloom goes unchecked, with the inevitable terrible consequences.

In 1995, researchers in the English Lake District demonstrated that the addition of small amounts of sodium phosphate to Seathwaite Tarn, a lake whose waters had become acidic, could increase biological productivity and restore the pH of the lake to pre-acidification levels. The work was a joint effort by William Davison and Neil Edwards of Lancaster University, and Glen George of the Institute of Freshwater Ecology, Ambleside. Over a period of two years the pH increased from 5 to 6 as the phosphate stimulated plant growth and phytoplankton increased ten-fold, which in turn used up much of the nitrate in the water that had been responsible for the acidity. Overall, each phosphate ion removed sixteen acidic ions. By the end of the experiment, the levels of micro-organisms and crustaceans in the lake had increased markedly and the brown trout population had also risen.

Reclaiming phosphorus from sewage

The worries about phosphate in natural waters, however, prompted people to take a closer look at ways of removing it from sewage, and there is every reason to believe that phosphate derived from sewage will be a renewable resource of the future.

The average human excretes around 2 grams of phosphorus a day in the form of 6 grams of phosphate, and almost all of it goes to waste. Human sewage and animal waste represent an important source of phosphorus. The responsibility for garnering phosphorus from the former falls to sewage-treatment plants, and in certain areas they are already required by law to remove it from waters discharged into the environment. However, while they are geared for this function, they have yet to find a way of gathering the phosphate so that it can be reused.

Sewage works can do one of two things to phosphate: precipitate it as a metal salt or encourage microbes to absorb it. Either way it ends up in the sludge which settles out from the sewage. Recovery of phosphorus from the sludge may one day be economically viable, or even enforced by law, and is to be encouraged. The recovered phosphate could have a definite saleable value to offset the cost of the process. Although natural deposits of high-grade phosphate ore are still relatively abundant, they are being used up at a rapid rate and a time will come when poorer-quality rock will have to be mined. At such a time, it may well be that phosphate reclaimed from sewage will come into its own. Currently, animal wastes, such as those from poultry houses and piggeries, could be a more attractive source from which to recover phosphate.

In 1998, the Centre Européen d'Etudes des Polyphosphates commissioned a report on phosphate recovery from animal manure in the Netherlands, which came to the conclusion that it would be viable. The report found that animal manure produced annually contains 86,000 tons of phosphorus, of which 92.5 per cent is returned to the land, leaving 7.5 per cent surplus to requirements, a proportion that is destined to rise to around 10 per cent by 2002. Farmers are unwilling to pay the costs of processing this surplus to recover the phosphorus, although they currently have to pay the cost of disposal. Moreover, financiers aren't willing to fund processing projects because some in the past have been commercial failures and local residents object strongly to processing plants being built in their neighbourhood.

However, large intensive livestock farms of the future could easily be linked with phosphorus-recovery plants, especially those of the veal calf industry. One plant in the Netherlands is already processing 100,000 cubic metres per year of fluid slurry and experimenting with extraction of the phosphorus as struvite (ammonium magnesium phosphate), which can be used as a slow-release fertilizer. The manure from pig farms is less easy to process, and a centralized facility for processing this went bankrupt in the early 1990s.

Chicken manure, on the other hand, has both a high phosphate content and a high energy content, and when this is used as fuel it can generate electricity and yields an ash that can be up to 25 per cent phosphate. Three such plants are already in operation in the UK burning 700,000 tons of chicken manure per year, which is two-thirds of all such waste.

Human manure is less easy to process because by the time it is collected it has been greatly diluted with water. It is also problematic when used as a fertilizer for three main reasons: first, the sewage can be contaminated with human pathogens; second, it might be contaminated with heavy metals, especially if it has come from an industrial area; and third, farmers do not like the idea of working with it. Even when the sewage has been irradiated to ensure that all disease organisms have been killed or when it has come from a non-industrialized area, farmers and other country-dwellers still object to its use.

Despite this, there is a growing need to find a valid use for the phosphate that will by law need to be removed from sewage in Europe. In 1991, the European Community issued its Urban Wastewater Treatment Directive regarding permitted levels of nitrate and phosphate in what are termed 'sensitive waters'. The directive had to be complied with by the year 2000. For sewage works that serve communities of fewer than 100,000 people, there must be no more than 2 mg of phosphate per litre and no more than 15 mg of nitrate in the exit waters, while for communities serving cities of more than 100,000, the conditions are more stringent: no more than 1 mg of phosphate per litre and no more than 10 mg of nitrate per litre.

Phosphate can be precipitated as an insoluble metal salt and for this the preferred metals are iron (added as ferric chloride or ferric sulphate), aluminium (added as aluminium sulphate or aluminium chlorohydrate), or calcium (added as lime). Aluminium precipitates aluminium phosphate in waters that are slightly acidic, with pH around 6, while iron precipitates iron(III) phosphate best in more acidic waters with a pH around 5 and calcium precipitates hydroxyapatite in alkaline waters, with pH above 8. The precipitation can be carried out at the first stage of sewage treatment, more or less as the sewage arrives at the works after it has been screened to remove gross items of detritus. Such treatment will generally only precipitate the simple phosphate and not that which is present as tripolyphosphate, so phosphate removal is of the order of 70–90 per cent. Some sewage works achieve remarkable results with primary-stage treatment alone. For example, the Hunt Creek sewage works in Virginia uses ferric chloride and reduces an influent water of 9.3 mg of phosphate per litre to 0.2 mg per litre in the effluent water.

Secondary precipitation is better and in this the chemical is added immediately prior to aeration so that the phosphate precipitate becomes part of the sludge in the aeration tank, with a volume increase of around 30 per cent. This secondary phosphate treatment can remove 80–95 per cent.*

Biological phosphorus-removal was first suggested as long ago as 1955 when it was discovered that activated sludge could take up phosphorus at much higher levels than bacterial growth would suggest possible. Vigorous aeration of activated sludge can cause the concentration of soluble phosphorus to decrease to less than 1 mg per litre. This early research led to the first biological phosphorus-removal process, known as the Phostrip process. Moreover, the sludge that settled from such treatment was suitable for use as an agricultural fertilizer.

In these systems the growth of certain biological strains that accumulate polyphosphates is promoted, provided the bacteria can scavenge volatile fatty acids necessary for growth, such as acetic acid. These are naturally present in sewage, especially in industrial areas. The bacteria with an appetite for phosphate are the *Acenitobacter* spp., *Aeromonas* and *Pseudomas* spp., which all happily digest raw sewage. Indeed, there are several commercial systems of biological phosphorus-removal now in use, such as the Bardenpho and A/O process, which were developed in the USA. The first of these is capable of reducing phosphorus by as much as 95 per cent and the latter by 80 per cent. Biological nutrient removal is carried out at 370 plants in the USA, and the phosphorus present in the anaerobic sludge-digesters ends up as struvite. Other processes need the addition of calcium to produce hydroxyapatite as the by-product. Having concentrated the phosphate thus, there is nothing, in theory, to stop its being used as a raw material for either the thermal or wet-acid process. Indeed, it should find a ready market because it is more pure than natural phosphate rock, which has high levels of fluoride and unwanted metals, such as uranium, cadmium, arsenic.

---

* Tertiary precipitation can be undertaken on the effluent from the secondary stage of treatment, but it is not often done.

A great deal of research has also gone into the ways of turning sewage sludge into a form of phosphorus that will be acceptable as fertilizer. Where iron and aluminium have been used to precipitate the phosphate, the product is not really suitable because the phosphorus is too tightly bound. It is currently too costly to turn this into the more plant-friendly calcium or ammonium hydrogen phosphate, which would make good fertilizer, but one day even this might be commercially viable. There are other ways of processing sewage to produce fertilizer phosphate, and this is easily achieved using the material derived from biological phosphorus-recovery methods.

Simon Environmental Engineering in the USA has developed the Simon-N-Viro process in which the sludge is mixed with cement-kiln dust. The mixture is heated at 52°C for twelve hours to sanitize it and is then left for about a week to compost, before being bagged for sale. The material is germ-free and any heavy metals are converted to insoluble hydroxides by the action of the cement dust. At La Rochelle in France, sewage sludge is simply dried, using heat from a municipal incinerator, and then pressed into pellets. Another process, known as the Swiss Combi Drier, uses a mixture of steam and air at 450°C to dry the sludge, which is rendered sterile. The process is also energy efficient because combustible gas and dust are burned to heat the water. One of the first Swiss Combi Drier process plants went into operation at Biel near Zurich in 1989 and produces 3,000 tons a year of solids, which are used as soil conditioner and fertilizer.

As yet the recycling of phosphate is in its infancy, but research has already shown that it is feasible. The day is fast approaching when the phosphate we ate for our dinner returns as the phosphate we put into our dishwasher, which returns as the phosphate we wash our clothes with, which returns as the phosphate we clean our teeth with, which returns as the phosphate in the cola we drink . . . and so on *ad infinitum*.

14. Spontaneous human combustion and other horrors

They advance slowly, look at all these things. The cat remains where they found her, still snarling at the something on the ground, before the fire and between the two chairs. What is it? Hold up the light!

Here is a small burnt patch of flooring; here is the tinder from a little bundle of burnt paper . . . and here is – is it the cinder of a small charred and broken log of wood sprinkled with white ashes, or is it coal? O Horror, he *is* here! And this, from which we run away, striking out the light and overturning one another into the street, is all that represent him.

Help, help, help! Come into this house for Heaven's sake!

Plenty will come but none can help . . . Call the death by any name . . . say it might have been prevented how you will, it is the same death eternally – inborn, inbred, engendered in the corrupted humours of the vicious body itself – spontaneous combustion – and none other, of all the deaths that can be died.

The scene Charles Dickens describes in *Bleak House* is one of spontaneous human combustion – in this case of the second-hand dealer and rogue Mr Krook, and when published in 1853 it brought the phenomenon to the attention of a mass audience. Moreover, it fuelled a debate which had been going on for more than a century and which continues to this day.

Most scientists and fire experts refuse to accept that spontaneous human combustion can happen, but there are those who
believe it is a genuine phenomenon and they point to plenty of
well-documented cases which defy other explanation. In Charles
Dickens' time there appeared to be good *scientific* grounds for
accepting that it could happen, and scientists believed it was
caused by a spontaneously flammable gas called 'phosphuretted
hydrogen'. If this were to be produced in the human gut and
passed as wind, it would immediately ignite, setting fire to that
person's clothing. If they were unconscious or in a drunken
stupor and incapable of taking action, then the outcome might
be just as Dickens described.

Dickens alludes to this somewhat embarrassing theory in his
account of Krook's inquest:

> Out of the court and a long way out of it, there is consider-
> able excitement too; for men of science and philosophy
> come to look and carriages set down doctors at the corner
> who arrive with the same intent and there is more learned
> talk about inflammable gases and phosphuretted hydrogen
> than the court ever imagined. Some of these authorities (of
> course the wisest) hold with indignation that the deceased
> has no business to die in the alleged manner; and being
> reminded by other authorities of a certain inquiry into the
> evidence for such deaths, reprinted in the sixth volume of
> the *Philosophical Transactions* . . . and likewise of the Italian
> case of the Countess Cornelia Baudi as set forth in detail by
> one Bianchini, prebendary of Verona, who wrote a scholarly
> work or so and was occasionally heard of in his time as
> having gleams of reason in him; and also of the testimony
> of Messrs Foderé and Mere, two pestilent Frenchmen who
> *would* investigate the subject; and further, of the corrobora-
> tive testimony of Monsieur Le Cat, a rather celebrated
> French surgeon, who had the unpoliteness to live in a house
> where such a case occurred and even to write an account of

it. Still they regard the late Mr Krook's obstinacy, in going out of the world by such a byway, as wholly unjustifiable and personally offensive.

It was with such gentle sarcasm and references to well-documented cases that Dickens sought to confront those who would not accept the possibility of spontaneous human combustion, and he mentioned 'phosphuretted hydrogen' as the explanation. This was the name given to a gas which had first been made in 1783 by a chemist called Gengembre, who heated phosphorus in caustic potash (potassium hydroxide). As the gas escaped from his apparatus it immediately caught fire and, while the gas is predominantly phosphane,* which is not of itself spontaneously flammable, it also contains a little of a second gas, diphosphane,† which is. It was the diphosphane which caused the gas emerging from Gengembre's reaction vessel to burst into flames.‡

It was only realized sixty years later that diphosphane was the key to flammability, when a French chemist named Paul Thénard obtained almost pure diphosphane by reacting calcium phosphide (Ca_3P_2) with water. This chemical reaction was to be made use of in self-igniting flares (see box overleaf).

If Dickens had been aware of Thénard's discovery he might have realized that the self-igniting of humans by diphosphane would be highly unlikely. Although PH_3 might be produced by decaying organic matter in the human gut, it was highly unlikely that P_2H_4 would also be produced. However, 150 years on, we now have evidence that, improbable though it may seem,

---

* Also known as phosphine, chemical formula PH_3.

† Also known as diphosphine, chemical formula P_2H_4.

‡ The diphosphane can be removed by passing the vapours through a flask immersed in a freezing mixture when it condenses out of the gas stream, and then the emerging vapour does not instantly burst into flames.

Self-igniting flares

In 1876 Nathaniel John Homes, a telegraph engineer living in Primrose Hill on the outskirts of London, was granted a patent for 'Improvements in self-igniting and inextinguishable signal lights for marine and other purposes'. In the years that followed these were to take the form of rescue lights for lifeboats, distress lights that could be hoisted to the top of the mast of a ship that was in trouble and signal lights dropped by aircraft to pinpoint locations at sea. They consisted of a mixture of calcium carbide and calcium phosphide. The carbide reacted with water to form acetylene gas (also known as ethyne gas), which burned with a bright light, and the phosphide reacted to form diphosphane, which ignited it. Once water came into contact with the contents of the signal it started to burn and nothing could extinguish it.

Distress rockets are now an essential part of all life-saving equipment used at sea, although modern self-igniting flares rely on magnesium phosphide rather than calcium phosphide to generate diphosphane.

---

diphosphane can be produced in the human gut. Spontaneous human combustion might just possibly occur – at least in theory.

In Dickens' day the issue was hotly debated, and it is instructive to examine the evidence he cited. Other writers had introduced the theme of spontaneous human combustion into their novels before Dickens, including Captain Frederick Marryat, author of *Children of the New Forest*, who had a similar scene in his book *Jacob Faithful*. The Jacob of the title finds his mother totally consumed by fire on her barge on the canal, although nothing else on the boat has been burned.

Dickens used the inquest into Krook's death to hint at a

scientific reason for how it could have happened – he could do little more than hint because the subject was considered indelicate – but we know from his preface to the book that he had researched the subject before he put pen to paper. Of the cases of spontaneous human combustion he referred to, the best known is that of the sixty-two-year-old Countess Cornelia Baudi of Cesena, Italy, who died in 1731.

On the evening of her death the Countess decided to go to bed early because she was feeling unwell. Her maid helped her prepare for the night and chatted with her for quite some time before leaving her alone. The following morning when the maid entered the bed chamber, all she found was a pile of ashes with only the countess's head and lower legs still recognizable. Perhaps because it involved a member of the aristocracy, this strange death attracted much public interest, not least among the scientists of the day. An account of the tragedy was even published in the leading scientific journal, the *Philosophical Transactions of the Royal Society*, in 1745. Paul Rolli, a fellow of the Society, had translated into English an account of the incident by Joseph Bianchini, who had written a long report on the affair. Part of Rolli's translation read as follows:

> Four feet distance from the bed there was a heap of ashes, two legs untouched, from the foot to the knee, with their stockings on; between them was the lady's head: whose brains, half of the back part of the skull and the whole chin, were burnt to ashes; among which was found three fingers blackened. All the rest was ashes, which had this peculiar quality, that they left in the hand, when taken up, a greasy stinking moisture.

The account went on to describe how the bedroom was covered in soot but nothing else had been burned, although candles in the room had melted. The bed was not damaged and was just as the countess had left it. The report mentioned a small oil lamp on the floor but noted that there was no oil in it. This could well be the key to the mystery. Was this left burning low in case the countess wanted to get up in the night? In which case, there is no need to postulate *spontaneous* human combustion, because a source of ignition was at hand. If the countess had got out of bed, had a heart attack and fallen so that her night gown was ignited by the lamp, then by morning she could indeed have been reduced to a heap of ashes. No raging inferno is needed to consume a body in this way, only the slow release of combustible body fat, as research in recent years has shown. The combustion is a slow process, taking many hours, but it can result in almost all of the organic material of the body being burnt, leaving only bone ash.

Tests, using a pig carcass wrapped in cloth, demonstrated that a body can burn over a period of several hours and reduce itself to ashes. Once the cloth ignites it will act as a kind of external wick, using fat rendered from the body as a supply of fuel. The heat generated is enough to melt the layer of fat beneath the skin and keep the fire burning, despite the fact that the body may be 60 per cent water. Eventually the fire transmits itself to the inner parts of the carcass, where intense local heating may occur, enough to calcine the bones of the skeleton to a fine ash.

All this can take place without sufficient heat being transferred to the room to cause other items, such as furniture and fabrics, to catch fire. A combination of fatal heart attack and a source of ignition such as a cigarette, a candle or an open fire will explain the majority of cases of 'spontaneous' human combustion. It also explains why most of the victims who die in this way are elderly and overweight.

Another of the incidents that Dickens mentions in *Bleak House* took place on 19 February 1725 at an inn in Rheims, where an apprentice surgeon, Claude Nicolas Le Cat, was staying the night. The charred remains of the innkeeper's wife, Mme

Millet, were discovered in the early hours of the morning and, again, nothing else in the room was burned. Jean Millet was arrested and charged with his wife's murder – he was thought to be having an affair with one of the serving girls, which provided him with a motive. Le Cat testified on his behalf and argued successfully that Mme Millet was a victim of spontaneous human combustion. She was known to be a heavy drinker and her cremated remains were found close to a fireplace, which probably means that it was not spontaneous. Perhaps she fell on the floor in a drunken stupor and as she did her clothes touched the fire and started to smoulder and burn; she would probably have been too far gone to do anything about it until it was too late.

In Charles Dickens' day, the popular explanation given for spontaneous human combustion was alcohol consumption. Indeed, those who campaigned against the demon drink quoted such cases as a warning of what might befall you. Justus von Liebig, whom we met in Chapter 12, took a scientific interest in the subject and wondered if drink really could be the cause. In 1851, he analysed fifty cases of spontaneous human combustion and came to the conclusion that it was highly unlikely alcohol was to blame. Liebig even carried out experiments to prove his point. When he ignited the flesh of anatomical specimens which had been preserved in alcohol, he found that the alcohol burned away, while the flesh was only slightly singed. Tests on rats injected with alcohol showed that they too were non-combustible. It may well be that Dickens was aware of Liebig's conclusions, which is why he supported the theory that spontaneous human combustion really was spontaneous and came from within, from 'phosphuretted hydrogen'. What caused this to form was something he did not care to speculate on.

At this point, I should declare that I am highly sceptical about *spontaneous* human combustion, but it is a subject that refuses to go away and the evidence sometimes seems highly compelling. In most cases when it has been reputed to occur, there has been a source of ignition near by and this would appear to account for the vast majority of cases, which are really *ignited* human combustion. Whether the cause is spontaneous or ignited human combustion, the outcome is the same and such deaths really do appear mysterious: a human body reduced to fine ash, with the possible exception of the head and the lower legs and feet.

It should not happen but it does

Even though phosphorus was taken as a medicament in the nineteenth century, it is only capable of producing phosphane and diphosphane when it comes into contact with strong alkali, such as potassium hydroxide. However, conditions in the human body are acid and the conditions in the stomach are strongly acid, so a chemical process for diphosphane formation from elemental phosphorus is impossible.

If these gases were to form, they would need to come from the plentiful supply of phosphate in the body or, more likely, the food in the gut. After all, the conditions in the intestines are anaerobic, and microbes there produce other hydrides, such as methane (CH_4) and even hydrogen gas itself (H_2). Might these same microbes be able to make PH_3 and P_2H_4 as well? If so, the emerging gas from the lower intestine might well be able to catch fire. Could it possibly happen? The answer from chemists had been a resounding 'no' – until recently.

It has always been the received wisdom of chemistry that there was no way in which phosphate (PO_4^{3-}) could naturally be changed to phosphane (PH_3). Such a change defied the laws of oxidation-reduction reactions, as well as going against the energy difference between these two species. Nothing in the natural world had the power to replace four, immensely strong phosphorus–oxygen bonds with three feeble phosphorus–hydrogen bonds. Even in the laboratory there was nothing that chemists had come across that was able to accomplish this conversion in a single step, and they came to believe that it must be impossible. In any case, such large quantities of energy would be required to make it happen that no natural system could have evolved that had the power to carry it out.

These assertions made it difficult to explain the phenomenon of Will-o'-the-wisp, those curious lights that flit across marshes at night. The early chemists had collected marsh gas that bubbled to the surface from bogs and stagnant waters and had shown that it was methane (CH_4). This would provide the fuel for the lights, but what caused the methane to ignite? Spontaneously flammable phosphuretted hydrogen seemed to provide the answer. The anaerobic conditions of a marsh that turned decaying organic matter to CH_4 could clearly convert phosphate to phosphuretted hydrogen. It seemed a logical explanation, but later chemists declared it impossible.

In fact it was the chemists who were wrong. Nature may be governed by the laws of chemical thermodynamics, but she often seems to ignore them. The production of phosphane and diphosphane is a case in point. Despite the received wisdom that it was impossible to produce phosphane in natural systems, some chemists reported that they had detected this gas in sewage sludge and in the sediment at the bottom of harbours. While this was puzzling, it was not entirely inexplicable: it might have come from industrial waste. However, the natural formation of *diphosphane* with its phosphorus-to-phosphorus bond was nigh on impossible, surely? Microbes knew otherwise.

In 1993, Günter Gassmann and Dieter Glindemann, of the Helgoland Biological Institute in Hamburg, discovered that micro-organisms could make phosphane and diphosphane after all. They published their research in the leading European journal of chemistry, *Angewandte Chemie, International Edition in English*. Gassmann and Glindemann concentrated on the anaerobic natural environment that exists within the gut of cows, where copious amounts of methane gas are formed.

In the intestines of freshly slaughtered animals they found what they were looking for: phosphane. They didn't find much, but it was present to the extent of 3 nanograms per kilogram (ng/kg) of material in the rumen of cattle and at a level of 9 ng/kg in fresh cowpats. When they turned their attention to pigs they found even more: 103 ng/kg in the colon and as much as 960 ng/kg in pig manure. This is still a tiny amount; a nanogram (ng) is a billionth of a gram.* Even so, some rare enzyme had apparently achieved the impossible: it was able to convert phosphate to phosphane.

Could this reaction occur in human intestines? We eat a phosphate-rich diet of foods and, as we saw in Chapter 13, these can have their phosphate levels boosted, for example processed meat, sausages, cheese spread and colas. If farm animals can produce phosphane in their guts, so should humans. Gassmann and Glindemann analysed human faeces and there it was: 160 ng/kg in those from an eight-month-old baby and 80 ng/kg in those collected from adults. Vegetarians produce much less – around 20 ng/kg – because they eat less phosphate-containing food. When a sample of liquidized human faeces was added to a culture medium and incubated under anaerobic conditions, Gassmann and Glindemann discovered that not only was more phosphane produced but *diphosphane* as well. After two weeks this had reached a level of almost 200 ng/kg.

Their research does not answer the question of why phosphane is produced, or how, but clearly there is a microbe with the necessary combination of enzymes to effect this remarkable transformation. However, their research did explain the Will-o'-

---

* A quantity as small as a nanogram (10^{-9}g) is difficult to imagine, but if we were talking about time rather than weight, then a billionth (10^{-9}) would be a second compared with thirty years.

the-wisp phenomenon. The scientific name for this is *ignis fatuus*. It might even explain ghosts, especially where a cemetery is located on water-logged land. It is not impossible that graves and coffins would fill with water and the putrefaction would generate methane along with phosphane and diphosphane. The reports of ghostly apparitions emerging from graves was capable of rational explanation after all.

Of course this is still only speculation, but it does make sense of some supernatural events. It did not, however, provide an answer for spontaneous human combustion. After all, diphosphane was detected only when human faeces were left to ferment for a week or two.

The link between diphosphane and spontaneous human combustion is tenuous, but at least it can explain how the human body might generate a chemical capable of igniting itself. A person who is badly constipated might have conditions within their intestines that would generate it, along with the methane gas and hydrogen that we know constitutes human wind. Diphosphane bursts into flame when it comes into contact with the oxygen of air and, if it ignited the expelled methane gas, it might well start a fire. At least in theory.

Typical cases of spontaneous human combustion

There are many cases of humans being mysteriously incinerated, but these are almost certainly cases in which the person died before the fire started and the body was ignited by an external flame. For it to be termed spontaneous combustion, all sources of accidental ignition must be eliminated and this is often not possible. In their book, *Spontaneous Human Combustion*, published in 1992, Jenny Randles and Peter Hough list 120 examples of the phenomenon, spanning almost 400 years. The first of these happened at a village near Christchurch in Dorset on 26 June

1613, when a carpenter was reduced to ashes in mysterious circumstances. Accounts of the event are vague and we have no way of knowing whether this was spontaneous human combustion or merely human combustion.

The *Philosophical Transactions of the Royal Society* of June 1744 carried details of a similar case, that of Grace Pett, who lived at St Clement, Ipswich. When members of her family, completely unaware that anything unusual had transpired during the night, went into Grace's bedroom at six o'clock on the morning of 9 April they found only her head and feet – the rest of her body had been reduced to ashes. Strangely the wooden floor on which these rested was unburned and, although there was a fireplace in the bedroom, it was said that there had been no fire lit there the evening before. The family and neighbours suspected witchcraft, but the coroner's jury prudently recorded a verdict of accidental death. Meanwhile, the fellows of the Royal Society talked of spontaneous human combustion.

On 2 March 1773 at Coventry, England, fifty-two-year-old Mary Clues of Gosford Street was discovered as a pile of ashes on the floor by the side of her bed (which was undamaged), and supernatural causes were again suspected. She had previously told her neighbours that the Devil had appeared to her one night saying he was coming to take her away. Others attributed her end to a fondness for drink. We are not told whether there might have been a source of ignition, such as an open fire or a candle, in the room where she was found.

On 12 May 1890, a Dr B. H. Hartwell of Ayer, Massachusetts, was called to the home of a forty-nine-year-old woman by her daughter, who said that her mother had been clearing stumps and roots from the wood behind their house when she had burst into flames. In the wood the doctor discovered the woman's body arched above the ground, which he said was due to the effect of the fire on her muscles, and noted she was burning like a log. The fire was extinguished by smothering it with soil.

Hartwell wrote an account of the incident for the *Boston Medical and Surgical Journal* in which he says the victim had lit a bonfire in order to burn the roots she was digging out of the ground.

The doctor thought this was not the cause of her catching fire because, he says, the burning body was about 30 feet (10 metres) from the burning pile of roots. Moreover the ground was damp from recent rain and it was unlikely that smouldering leaves on the ground could have set her clothes alight. However, it is not difficult to devise a scenario that could account for the incident without invoking spontaneous human combustion: perhaps her clothes accidentally caught fire and she ran to find help but passed out and suffocated. If she had remained undiscovered for some time, the wick effect would have begun and the body would have started to burn.

One of the most celebrated cases of spontaneous human combustion – celebrated because of the completeness of the incineration and the lack of damage to the surroundings – was that of plump (170 lb) sixty-seven-year-old Mrs Mary Reeser of St Petersburg, Florida, which happened on 1 July 1951. Mary lived in an apartment at 1200 Cherry Street where she was regularly visited by her son, Dr Richard Reeser, who often had morning coffee with her. That Sunday, Mary went to her son's home for lunch and did some baby-sitting but returned home mid-afternoon after a family quarrel. At eight o'clock that evening Richard went round to see his mother, found her ready for bed and talked with her for about an hour before going home. While he was with her she smoked a cigarette and had already taken two sleeping pills.

At eight the following morning a telegram arrived for Mary and the discovery was made: she and the armchair she had been sitting in had been reduced to a pile of ashes from which protruded an unburned left foot and its black satin slipper. Most of the things in her room were blackened but not burned. Although this case is famous among the spontaneous human combustion fraternity, it is more than likely that Mary lit another cigarette, fell into a deep sleep or had a heart attack and the smouldering cigarette then ignited her rayon nightdress.

It is now accepted that human bodies can provide enough fuel from the fat they contain to incinerate themselves completely, but in some cases this happens rather too quickly to fit the theory of slow, intense, localized combustion. In December 1956 Young Sik, a seventy-eight-year-old disabled man living in Honolulu, Hawaii, was discovered enveloped in a blue flame. His neighbour immediately summoned the fire brigade, who arrived within twenty-five minutes, by which time Sik's body was said to be completely consumed.

In September 1967 in Birmingham, England, the body of a middle-aged vagrant was found ablaze in a derelict house with flames coming from his stomach 'like a blow torch'. This was presumed to be alcohol burning, but an autopsy showed that the man had actually died of asphyxiation from the fumes of his own burning body, which meant he had remained alive for some time after the fire started. The cause of the fire was recorded as 'unknown'. Supplies of gas and electricity to the building had been disconnected and nothing in the way of cigarettes, matches or a lighter was found near the body. It was assumed that the man had been too drunk to put out the flames when they started and that possibly the undigested alcohol in his stomach had added fuel to the fire once his flesh had burned through.

Spontaneous combustion of the living

There are only a handful of cases where there has been no evidence of an external source of ignition. For example in October 1776 at Fenile, Italy, a priest was reputed to have burst into flames shortly after arriving at his sister's house. Those who ran to his aid found him writhing on the floor surrounded by flames, and the priest himself, who survived for several days, said it had started when a bluish flame suddenly issued from his shirt. The case was considered sufficiently curious to be reported in a medical journal.

On a bitterly cold day in January 1835, Professor James Hamilton of the University of Nashville, Tennessee, walked home from the university and paused before entering the house to take readings from a thermometer on the outside wall, noting that the temperature was 8°F (−13°C). Suddenly he felt an intense pain in his left thigh and, looking down, saw a bright flame burning through his trousers. With some difficulty he beat it out, then rushed indoors and took the trousers off, to discover a hole burned into his underpants and a deep wound on his leg. He went to see his doctor, John Overton, who was so intrigued by what had happened that he wrote an account of the incident in the *Transactions of the Medical Society of Tennessee*.

Also in the USA, in 1847, another medical man, Dr Nott, wrote an article for the magazine *Scientific American* in which he reported how a twenty-five-year-old man, 'an habitual drinker', was discovered standing in a silver-coloured flame which roasted him alive. The doctor attested that he had seen the man only two hours earlier and that there was no possibility the fire had been caused by an external source.

A somewhat similar eruption of fire from a living person was reported in *China Youth News* as recently as April 1990. The story concerned a four-year-old boy, Tong Tangjiang, whose trousers suddenly started emitting smoke and then began to burn. The boy was rushed to hospital where the same thing happened again, this time burning his bedclothes and mattress.

There have been some cases of people inexplicably starting to burn while lying in bed – and who have lived long enough to tell the tale.

On 23 December 1904 in Hull, a Mrs Clark, who lived as a pensioner at the Trinity Alms-house, was found alight in bed although the bed itself was hardly burned. She lived long enough to be questioned about what had happened but could give no explanation as to why the blaze had started.

Mrs Madge Knight of Chichester, Sussex, was lying in bed on the night of 6 December 1943 when suddenly her nightdress caught fire along her back. Her screams brought help and medical aid, but she died in the local hospital soon after, unable to account for what had happened.

A similar disaster befell a twenty-year-old student on 13 May 1980 in Birmingham. He awoke in the early hours of the morning to find himself in a sea of flames and his cries awoke fellow students in adjacent rooms. They put out the fire by smothering it with pillows, then carried the student into the corridor and called the fire brigade and an ambulance. Help arrived at 3.30 a.m. and, while firemen examined the room, paramedics administered first aid. He died in hospital two days later from 70 per cent burns, still unable to say what had happened. The firemen found nothing to explain the cause of the mysterious fire and the coroner eventually recorded the death as 'due to fire, cause unknown'.

Also difficult to explain is the phenomenon of clothes suddenly catching fire. Phyllis Newcombe of Chelmsford went with her fiancé, Henry McAusland, to a dance on 27 August 1938. As they were leaving the dance-floor her dress suddenly burst into flames and, despite the efforts of Henry to beat them out, Phyllis died within minutes. At the inquest, the coroner suggested that a cigarette might have ignited her dress, but when a similar dress was tested it was found impossible to ignite it with a cigarette.

On 28 January 1985, a teenager who was walking down a flight of stairs with a group of fellow students suddenly burst into flames, becoming what was described as 'a human torch'. The incident happened at the Halton College of Further Education in Widnes. One of the other students said it started as a small flame down her back. Although she soon appeared to be making a good recovery in hospital, she died two weeks later from lung damage. While she was in hospital she was questioned about the fire but was unable to say what had caused it. One explanation was that she brushed against a lit gas-ring in the cookery class she had been attending and that this had set her clothes smouldering. The draught of air up the stairwell then fanned this into flames. The coroner's jury returned a verdict of accidental death.

25 May 1985: Paul Hayes, a nineteen-year-old computer-operator, was walking alone along the street at Stepney Green in London when he suddenly found flames rising from his waist which set his clothes burning. He rolled on the ground to put the fire out, then staggered to a nearby hospital where he was treated for burns to his torso. He could offer no explanation as to how the fire had started.

3 October 1987: an eighty-six-year-old man was observed by his housekeeper suddenly to emit flames from his nose and mouth and then from his lower body. He quickly burned, along with the settee he was sitting on, but official investigations into his death could not account for the cause of the fire.

Spontaneous combustion of the dead has also been reported. In 1866 a man's corpse was found burning in a church vault, a year after his burial. He had been thirty when he died of typhoid. The day before the fire a foul smell had been noticed coming from the crypt; when this was investigated it was discovered that his coffin had split and a vile liquid was oozing from it. Sawdust was heaped over it to absorb the mess. The following morning when workmen went to remove the coffin, it was burning with a curious bluish flame.

This case has almost all the requirements for a fire started by diphosphane. Clearly the man's body was rotting under anaerobic conditions which produced a pressure of methane gas, enough to burst the coffin. There would also have been the offensive sulphur gases which such conditions also generate. If phosphane were also produced this would account for the unique awfulness of the smell and, if diphosphane were produced, then ignition might well occur which would set fire to the methane and then the sawdust.

△

Only one or two of the above cases can possibly be attributed to the effects of diphosphane. If methane is the gaseous fuel and diphosphane merely the spark that sets it off, then cows should be more at risk than humans because they produce prodigious quantities of methane in their guts. Outbreaks of spontaneous bovine ignition in cowsheds are so rare as probably to be relatively non-existent, but unless a cow was constipated it would be unlikely that the bacteria would have time to generate diphosphane in its gut. Gassmann and Glindemann's work shows that diphosphane would form but in such small amounts as to be an ineffective source of ignition. Perhaps it requires a decomposing body rich in phosphate in order for the conditions to be right to produce enough diphosphane to achieve ignition.

In *Spontaneous Human Combustion*, there is the story of a sheep suddenly bursting into flames. It occurred one night at Weymouth, Dorset, during the Second World War and was witnessed by an officer and two of his men who were patrolling the coastline. According to Raymond Reed, serving with the 9th Battalion of the Royal Welch Fusiliers:

> One night we were crossing open downland where there were lots of sheep grazing. It was pitch black. Suddenly, without warning, a fire erupted about a hundred yards away ... we approached carefully and discovered it was a sheep on fire. The animal was on its side. From its stomach area issued blue flames. We were absolutely astounded. The sheep was a large animal, in no way decomposed, in fact quite fresh looking. I think it was already dead but cannot be

certain. We extinguished the fire by throwing earth and clods on to it.

It has been known for centuries that farmers who gathered hay before it was dry were likely to find their haystack bursting into flames. Any rotting vegetable matter piled in a heap is capable of getting warm, as many compost heaps demonstrate. The fuel to feed such a release of energy is easily explained: it comes from the carbohydrate that makes up the bulk of plants. When this carbohydrate is digested by microbes, they release heat as well, which assists their activity for a time, although when the pile of rotting vegetation gets too hot they will be killed. What may be happening in such anaerobic conditions is that some microbes are converting the carbohydrate into methane gas and this fuel pervades the heap. All that is then needed to start a fire is diphosphane to be generated, to diffuse to the outside and come in contact with oxygen.

While the effects of diphosphane may be disputed, its production by microbes still has to be explained. The impetus for carrying out research into its production would be to uncover the micro-organism and enzymes that are capable of achieving it. The chemical reduction of phosphate to phosphorus and then to phosphane has such high energy requirements that it is difficult to imagine it happening naturally. Even if we discovered Nature's secret process there would not be much call for an improved method of phosphane production, although there are some uses for this unusual gas (see box).

And what of spontaneous human combustion? There is no evidence to support the popular impression that fire can suddenly start inside the human body and, within a short time, completely incinerate it. Almost all cases in which a body ends up like this can be explained by an external source of ignition and the wick effect. It is just possible that one or two incidents might involve diphosphane generated by the body, but this seems

---

Spontaneous food preservation

The use of phosphane as a fumigant began in Germany in the 1930s when the chemical industry developed aluminium and magnesium phosphides as suitable agents for protecting food stored in warehouses or transport containers from insect pests. If tablets of these phosphides were left on trays, they would react with water vapour in the air and give off phosphane gas, which is highly poisonous to insects at quite low concentrations. The tablets would only react when the temperature of the warehouse was above 5°C (40°F). A handful of tablets would continue to provide protection for weeks and months. The only drawback was that some insect pests eventually built up resistance to phosphane and it became necessary to use the alternative fumigant, methyl bromide. This is a gas which requires cylinders and piping to be installed and while it is very effective it is reputed to be environmentally dangerous because of the damage it might cause to the Earth's ozone layer.

---

an unlikely scenario, except when the body is dead and decaying. Spontaneous human combustion, as such, is probably a myth. On the other hand, that equally mythical phenomenon, Will-o'-the-wisp, who darts with his lantern across the marsh at night, may well be genuine – and the same may be true of graveyard ghosts.

Epilogue: The Devil's element?

As this book has shown, the history of phosphorus is truly shocking – but why? After all it is only one of around ninety elements that are found on Earth and as such it is less harmful than many, such as mercury, lead and uranium. It occurs naturally only in the form of phosphate, PO_4^{3-}, which is about as safe as it's possible to be. Indeed this chemical grouping has some very useful features, such as the ability to link to other phosphates and to attach one or more organic groups to its oxygen atoms to produce materials such as DNA, which is the key to life itself.

However, when we strip away its protective cage of four oxygen atoms and expose the element itself, we release a tiger. Chemically this may not seem an easy thing to do because these oxygens are bound to phosphorus with some of the strongest of all chemical bonds, and there's the rub. In fact it is easy to do: fire will do it, as the alchemists discovered. All it needed was a mixture of phosphate and carbon and a very high temperature and it was bound to happen.

When something like wood burns the temperature it generates is not hot enough to cause this reaction to occur. What is needed is at least 1,000°C, which is just about attainable with a charcoal furnace, and then, given the right raw material, the element will come bounding out. The alchemists' fascination with urine provided them with an ideal combination of chemicals for generating phosphorus. And so it came to pass. An age that had no concept of hazards, safety checks or stringent medical tests suddenly had access to one of the most dangerous of materials.

The enigma of phosphorus lies in its chemistry. Because it is difficult to tear the element away from oxygen, we should not be surprised that these two elements fervently seek to recombine. In so doing they release the same amount of energy that was required to separate them, which is why burning phosphorus is so dangerous. Not only that, but the chemical reaction between the two begins immediately they come into contact.* The spontaneous flammability of phosphorus accounts for a great deal of its shocking history. Useful as that property appeared, attempts to utilize it for the benefit of humankind were always bound to be fraught with danger, as the story of the lucifer match illustrates. With care it was possible to make white phosphorus serviceable but that did not address its other deadly feature, its toxic nature.

When phosphorus is released from its cage of oxygens it does not then exist as single atoms, it straightway clusters together in groups of four, which is why the chemical formula for white phosphorus is P_4. By bonding to itself in this way it relieves some of its craving to form chemical bonds. The result is a pyramid of four atoms with three at the base and one on top, and this structure confers some stability on the molecule, at least to the extent that it is stable in water. Unfortunately, what this means is that when P_4 enters the body it can also survive long enough to reach the liver, the organ that recognizes it as alien, but whose attempts to dispose of it only means it wreaks its damage there. Weak though the molecule is, P_4 is both volatile and stable enough to survive long enough in the air to be breathed in – as those who worked in match factories were to discover.[†]

---

* Contrast the energy barrier to methane gas and oxygen reacting with each other. Although when they do react they give out incredible amounts of energy, their interaction has to be kick-started with an *input* of energy, in the form of a spark or flame. Not so with phosphorus.

† In the solid state the reaction of phosphorus and oxygen generates heat that cannot be dissipated, and this serves to melt it, thereby exposing more

Flammability and toxicity together account for all the damage and misery that elemental phosphorus has caused, whereas the damage attributed to human exploitation of phosphates springs from an entirely different set of properties. The overuse of phosphates contributed to environmental damage because this simply took the brake off this limiting factor in certain vulnerable locales. Then it was just too much of a good thing for unwanted species such as algae and pollution ensued. But this is a situation that has now been remedied and is not likely to occur again, given that future generations will recycle it from their sewage.

Modifying phosphate by attaching other chemical groups to the oxygen atoms, plus replacing one or more of these oxygens with sulphur or carbon, leads us into an entirely different world of benefits and evils. This can be the way to new healing drugs, but the effort which has gone into developing these is tiny compared to the amount of effort that has gone into finding molecules that can harm. While we may approve of chemical research that produces substances that will kill insects and other pests that threaten our food supplies, we can only condemn the enormous effort that has gone into making substances that will kill humans. The chemistry of organophosphates presents us, not with the dichotomy of either good or evil, but with a spectrum which ranges from the essential (DNA) at one end to the deadliest of agents (nerve gases) at the other. As we go from one extreme to the other we encounter healing drugs, useful household chemicals, herbicides and pesticides. To what extent these categories might overlap can be a cause for concern, but we must be assured that current regulations will ensure that only those that are completely safe will be allowed to be used.

---

phosphorus to attack by oxygen, which increases the temperature even more, and so on. In the gaseous state a molecule of P_4 will react with oxygen, once the two elements have collided, but this reaction is much slower and the energy is dissipated harmlessly.

So has the shocking history of phosphorus ended? Can we now look forward to a golden future utilizing the benefits this element can bring? I believe the answer is yes. It was because we stumbled upon this element long before we had any knowledge of chemistry that we allowed it too much freedom. Instead of 1669 being the year of its discovery we could have wished it had remained hidden for another 200 years, to a time when chemistry was firmly established. If it had been discovered in 1869, it would have been the sixty-third element to have been found rather than the thirteenth, but even so it would still have been too soon to prevent some of the havoc that most of this book has been about.

Whenever it was discovered, phosphorus would no doubt have been called upon to do the Devil's work. All we can hope for the new millennium is that that work is now complete.

Appendix: The right chemistry

FACT SHEET

PHOSPHORUS
[French, *phosphore*; German, *Phosphor*;
Italian, *fosforo*; Spanish, *fósforo*]

Chemical symbol P

Chemical properties
Element number 15
Atomic weight 31
Melting point 44°C (white); 590°C (red, under pressure)
Boiling point 280°C (white); 417°C (red, sublimes)
Density 1.82 kg per litre (white); 2.35 kg per litre (red)
Other forms see below
Radioactive isotope ^{32}P, half-life 14.2 days

Levels in humans
Blood 345 mg per litre
Bone 6.7–7.1 per cent
Liver 3–8.5 p.p.m.
Muscle 0.30–0.85 per cent
Average daily dietary intake 900–1,900 mg
Recommended daily intake 800 mg
Total mass of element in average (70 kg) person 840 g

Abundances (p.p.m.)*
In the Earth's crust 1,000 (PO_4^{3-})
In the oceans *Atlantic surface waters* 0.0015
Atlantic deep water 0.042
Pacific surface waters 0.0015
Pacific deep water 0.084
Residence time in the sea 100,000 years
Oxidation state in sea water +5 (PO_4^{3-}, HPO_4^{2-}, $H_2PO_4^-$)

* p.p.m. = parts per million

Allotropes of phosphorus

There are several structural forms (allotropes) of elemental phosphorus: white, red, Hittorf's and black being the best known. White phosphorus (also called yellow phophorus) consists of clusters of four phosphorus atoms in a pyramidal array (P_4), and this is the form made by the reduction of phosphate with carbon. When white phosphorus is heated under pressure at around 300°C for several days it changes to red phosphorus, which consists mainly of P_4 tetrahedra linking together to create a random network. In 1865 a German chemist called Johann Hittorf (1824–1914) dissolved phosphorus in molten lead and allowed it to cool, whereupon purple crystals of a new form of phosphorus were formed, and this consists of clusters of eight and nine phosphorus atoms linked to form a kind of tube of phosphorus atoms.

In 1916 an American chemist, Percy Bridgman (1882–1961), heated white phosphorus at 200°C under a pressure of 12,000 atmospheres and obtained black shiny crystals rather like graphite. This was black phosphorus, the most stable kind of phosphorus of all. The phosphorus atoms had arranged themselves into parallel layers, and like graphite this form, which it resembles, too was a semi-conductor of electricity. Other forms of black phosphorus, with different crystal shapes, can be made, depending on the pressure and the length of time it is heated.

Altogether about a dozen forms of phosphorus have been prepared ranging from crystal clear through all shades of orange, red, purple, brown and grey to deepest black.

Sources

1. Out of alchemy

K. Diem and C. Lentner, eds, *Scientific Tables*, 7th edn, J. R. Geigy, Basle, 1970 (phosphorus content of human body, etc.).

E. Farber, *The History of Phosphorus*, US National Museum Bulletin no. 240, 1966.

C. Gilchrist, *The Elements of Alchemy*, Element Books Ltd, Shaftesbury, 1991.

J. V. Golinski, *A Noble Spectacle: Phosphorus and the Public Cultures of Science in the Early Royal Society*, ISIS, vol. 80, 1989, p. 11.

A. L. Lavoisier, *Elements of Chemistry in a Systematic Order*, 3rd edn, trans. R. Kerr, Edinburgh, 1796.

K. Lewis, *The Chemical Works of Caspar Neumann MD*, London, 1759.

H. Muir, *Larousse Dictionary of Scientists*, Larousse, Edinburgh, 1994.

J. Murray, *A System of Chemistry*, Edinburgh, 1819.

J. R. Partingon, 'The Early History of Phosphorus', *Science Progress*, vol. 30, 1936, p. 402.

D. R. Peck, 'The History of Phosphorus and its Inorganic Compounds', PhD thesis, University of London, 1968.

R. L. Rawls, 'Dazzling Phosphorus', *Chemical & Engineering News*, 17 August 1998.

The Hermetic and Alchemical Writings of Paracelsus the Great, trans. A. Waite, James Elliot & Co., London, 1894.

M. E. Weeks and H. M. Leicester, *Discovery of the Elements*, 7th edn, ACS, Easton, PA, 1968.

2. The alchemist and his apprentice

R. Boyle, *A Short Memorial of some Observations made upon an Artificial Substance that Shines without any Precedent Illustration*, Hook's Lecture Cutler no. xi, London 1677.

R. Boyle, *New Experiments and Observations made upon the Icy Noctiluca*, London, 1681.

R. Boyle, *Philosophical Transactions of the Royal Society*, no. CXCVI, January 1692–3, p. 583.

F. F. Holmes, University of Kansas Medical Centre, private communication.

L. M. Principe, *The Aspiring Adept: Robert Boyle and his Alchemical Quest*, Princeton University Press, Princeton, New Jersey, 1998.

F. H. W. Sheppard, ed., *Survey of London*, vol. XXXVI, The Parish of St Paul, Covent Garden, Athlone Press, 1970.

3. The toxic tonic

F. E. Camps ed., *Gradwohl's Legal Medicine*, 2nd edn, John Wright & Son, Bristol, 1968.

P. Cooper, *Poisoning by Drugs and Chemicals, Plants and Animals*, 3rd edn, Alchemist Publications, London, 1974.

C. J. Polson and R. N. Tattersall, *Clinical Toxicology*, English Universities Press, London, 1959.

K. Simpson, ed., *Taylor's Principles and Practice of Medical Jurisprudence*, 12th edn, Churchill, London, 1965.

J. A. Thompson, *Free Phosphorus in Medicine*, H. K. Lewis, London, 1874.

K. A. Wade, ed., *Martindale: The Extra Pharmacopoeia*, 27th edn, The Pharmaceutical Press, London, 1977.

R. A. Witthaus, *Manual of Toxicology*, William Wood, New York, 1911.

A. C. Wootton, *Chronicles of Pharmacy*, Milford House, Boston, 1910; republished 1971.

4. Strike a light

W. A. Bone, 'The Centenary of the Friction Match', *Nature*, vol. 119, 1927, p. 495.

George Brown, *The Big Bang: A History of Explosives*, Sutton Publishing, Stroud, 1998.

M. F. Crass Jr, 'A History of the Match Industry', *Journal of Chemical Education*, Parts I and II, March 1941, pp. 116 and 118; Parts III and IV, June 1941, pp. 277 and 280; Parts V, VI and VII, July 1941, pp. 316, 317 and 318; Part VIII, August 1941, p. 380; and Part IX, September 1941, p. 428.

C. Finch and S. Ramachandran, *Matchmaking: Science, Technology and Manufacture*, Ellis Horwood, Chichester, 1983.

S. Jaffa, *Great Financial Scandals*, Robson Books, London, 1998.

W. T. O'Dea, *Making Fire*, Her Majesty's Stationery Office, London, 1964.

Hazel Rossotti, *Fire*, Oxford University Press, Oxford, 1993.

T. A. Watson, the history of matches told in two papers, *Manufacturing Chemist*, December 1936, p. 400; and January 1937, p. 23.

5. Strike!

C. A. Finch and S. Ramachandran, *Matchmaking: Science, Technology and Manufacture*, Ellis Horwood, Chichester, 1983.

Hackney Town Hall has the Bryant & May archive, although application to view needs approval of the company.

The London Hospital has an archive of records of phossy-jaw cases.

D. C. Mitchell, *The Darkest England Match Industry*, The Salvation Army, London, 1973.

E. Royston Pike, *Human Documents of the Age of the Forsytes*, George Allen & Unwin, London, 1969. This contains in full the Annie Besant articles she published in the *Link* and which led to the match girls' strike.

6. The cost of a box of matches

R. Beer, *The Match-girls Strike 1888*, Labour Museum Pamphlets, no. 2, London, undated.

H. Heimann, 'Chronic Phosphorus Poisoning', *Journal of Industrial Hygiene and Toxicology*, vol. 28, 1946, p. 142.

D. Hunter, *The Diseases of Occupations*, Hodder & Stoughton, London, 1976.

A. E. W. Miles, 'Phosphorus Necrosis of the Jaw: Phossy Jaw', *Journal of the British Dental Association*, 1973, p. 6.

S. Roalman, 'Matchgirls Ignite', *Spare Rib*, no. 73, p. 44, August 1978.

T. E. Thorpe, T. Oliver and G. Cunningham, *The Report on the Use of Phosphorus in the Manufacture of Lucifer Matches*, Her Majesty's Stationery Office, London, 1899.

7. Gomorrah

R. Barker, *The Thousand Plan*, Chatto & Windus, London, 1965.

M. Caidin, *The Night Hamburg Died*, Ballantine, New York, 1960.

V. Hodgson, *Few Eggs and No Oranges*, Dennis Dobson, London, 1976.

M. Middlebrook, *The Battle of Hamburg: Allied Bomber Forces against a German City in 1943*, Allen Lane, London, 1980.

G. Musgrove, *Operation Gomorrah: The Hamburg Firestorm Raids*, Jane's, London, 1981.

H. Rumpf, *The Bombing of Germany*, trans. Edward Fitzgerald, Frederick Muller, London, 1963.

C. Whiting, *The Three-Star Blitz; The Baedeker Raids and the Start of Total War 1942–1943*, Leo Cooper, London, 1987.

8. The ultimate evil – and a power for good

H. Bartle, 'Quiet sufferers of the silent spring', *New Scientist*, 18 May 1991, p. 30.

R. M. Black and G. S. Pearson, 'Unequivocal evidence', *Chemistry in Britain*, July 1993, p. 584.

J. Borkin, *The Crime and Punishment of IG Farben*, André Deutsch, London, 1979.

S. Budavari, ed., *The Merck Index*, 11th edn, Merck & Co., Rahway, NJ, 1989.

R. Chaudhry, S. B. Lall, B. Mishra and B. Dhawan, 'A foodborne outbreak of organophosphate poisoning', *British Medical Journal*, vol. 377, 1998, p. 268.

Crop Protection Reference, 11th edn, Chemicals & Pharmaceutical Press, New York, 1995.

J. Emsley and C. D. Hall, *The Chemistry of Phosphorus*, Harper & Row, London, 1976.

B. Hileman, 'Methyl Parathion: EPA's Challenge', *Chemical & Engineering News*, 27 January 1997, p. 22.

'Improvements in the manufacture of organic phosphorous compounds containing sulphur', British patents 1,346,409/1974, awarded to R. V. Ley and G. L. Sainsbury, and 1,346,410/1974, awarded to A. W. H. Wardrop and C. Stratford (VX agents), British Library Patent Office, London.

B. N. La Du, 'Structural and functional diversity of paraoxonases', *Nature Medicine*, vol. 2, no. 11, p. 1186, Nov. 1996.

T. C. Marrs, R. L. Maynard and F. R. Sidell, *Chemical Warfare Agents*, John Wiley & Sons, Chichester, 1997.

G. Matthews, *Pesticide Application Methods*, Blackwell, London, 1992.

M. Rudduck, 'Towards a safer world', *Chemistry in Britain*, March 1997, p. 25.

B. C. Saunders, *Some Aspects of the Chemistry and Toxic Action of Organic Compounds Containing Phosphorus and Fluorine*, Cambridge University Press, Cambridge, 1957.

A. Speer, *Inside the Third Reich*, trans. R. Winston and C. Winston, Weidenfeld & Nicolson, London, 1970.

T. F. Watkins, J. C. Cackett and R. G. Hall, *Chemical Warfare, Pyrotechnics and the Fireworks Industry*, Pergamon Press, Oxford, 1968.

Various authors, *Chemistry in Britain*, July 1988, special edition devoted to chemical warfare agents.

9. Murder

I. Butler, *Murderer's England*, Robert Hale, London, 1973.

J. Glaister, *The Power of Poison*, Christopher Johnson, London, 1954.

J. Rowland, *Poisoner in the Dock*, Arco, London, 1960.

K. Simpson, ed., *Taylor's Principles and Practice of Medical Jurisprudence*, 12th edn, vol. II, Churchill, London, 1965.

C. Wilson and P. Pitman, *Encyclopaedia of Murder*, Arthur Barker, London, 1961.

News Chronicle and *Daily Mirror*, 1 August 1953; *Daily Telegraph*, 7 August 1953 (the Merrifield case, trial reports).

10. Fortunes from phosphorus

J. Emsley and C. D. Hall, *The Chemistry of Phosphorus*, Harper & Row, London, 1976.

J. B. B. Lagrange, *A Manual of a Course of Chemistry*, London, 1800.

R. E. Threlfall, *100 Years of Phosphorus Making 1851–1951*, Albright & Wilson, Oldbury, 1951.

A. D. F. Toy, *Phosphorus Chemistry in Everyday Living*, American Chemical Society, Washington, DC, 1976.

11. Unlucky days

'Pyrophoric cargo BLEVE', *Hazardous Cargo Bulletin*, January 1984, p. 24.

'Hazardous materials fire', *Chemistry & Engineering News*, 21 July 1986, p. 4.

'Large-scale evacuation in Ohio', *Hazardous Cargo Bulletin*, October 1986, p. 75.

'Two fires in two hot weeks!', *Chemistry & Industry*, 20 August 1990, p. 504.

G. L. Fletcher, 'The dynamics of yellow phosphorus in Atlantic cod and Atlantic salmon; biological half-times, uptake rates and distribution in tissues', *Environ. Physiol. Biochem.*, vol. 4, 1974, pp. 121–38.

P. M. Jangaard, ed., *Effects of Elemental Phosphorus on Marine Life*, Fisheries Research Board of Canada, Halifax, Nova Scotia, 1972 (twenty-eight collected papers on the Placentia Bay disaster).

12. The supreme ruler

I. Asimov, *Asimov on Chemistry*, Macdonald & Jane's, London, 1975.

J. Emsley, 'Phosphate Cycles', *Chemistry in Britain*, December 1977, p. 459.

J. Emsley, 'The Phosphorus Cycle', in O. Huntzinger, ed., *The Handbook of Environmental Chemistry*, vol. 1/Part A, Springer-Verlag, Berlin, 1980.

E. J. Griffith, A. Beeton, J. M. Spencer and D. T. Mitchell, *Environmental Phosphorus Handbook*, Wiley & Sons, New York, 1973.

J. Kutsky, *Handbook of Vitamins, Minerals and Hormones*, 2nd edn, Van Nostrand Reinhold, New York, 1981.

U. Pierrou, 'The Global Phosphorus Cycle', SCOPE report 7, *Ecological Bulletin*, Stockholm, 1976.

13. Oh, shit!

G. R. Alexander Jr and D. A. Wallgren, *Detergent Phosphate Ban*, US Environmental Protection Agency, Region 5, Chicago, 1977.

A. M. Beeton, 'Eutrophication of the St. Lawrence Great Lakes', *Limnology and Oceanography*, vol. 10, 1965, p. 240.

S. Brett, J. Guy, G. K. Morse and J. N. Lester, *Phosphorus Removal and Recovery Technologies*, Environmental and Water Resource Engineering Section, Imperial College, London, 1997.

C. T. P. Coultate, *Food, the Chemistry of its Components*, 3rd edn, Royal Society of Chemistry, London, 1996.

J. Driver, 'Phosphates recovery for recycling from sewage and animal wastes', *Phosphorus and Potassium*, issue no. 216, July/August 1998, pp. 17–21.

G. E. Hutchinson, 'Eutrophication', *American Scientist*, vol. 61, 1973, p. 269.

Phosphate, Centre Européen d'Etudes des Polyphosphates, Brussels, 1997.

Phosphates: A Sustainable Future in Recycling, Centre Européen d'Etudes des Polyphosphates, Brussels, 1998.

Arthur D. F. Toy, *Phosphorus Chemistry in Everyday Living*, American Chemical Society, Washington, DC, 1976.

B. Wilson and B. Jones, eds, *The Phosphate Report*, Landbank Environmental Research & Consulting, London, 1994.

B. Wilson and B. Jones, eds, *The Swedish Phosphate Report*, Landbank Environmental Research & Consulting, London, 1995.

14. Spontaneous human conbustion and other horrors

M. Benecke, 'Spontaneous Human Combustion – Thoughts of a Forensic Biologist', *Skeptical Inquirer*, vol. 22, March/April 1998, p. 47.

J. Garratt, 'Will-o'-the-wisps and hot compost', *Chemistry Review*, vol. 3, March 1994, p. 30.

G. Gassmann and D. Glindemann, 'Phosphane (PH_3) in the biosphere', *Angewandte Chemie, International Edition in English*, vol. 32, 1993, p. 761.

R. E. Guiley, *Guinness Encyclopedia of Ghosts and Spirits*, Enfield, Middlesex, 1992.

D. Halliday, A. H. Harris and R. W. D. Taylor, 'Recent Developments in the Use of Phosphine as a Fumigant for Grains and Other Durable Agricultural Products', *Chemistry & Industry*, 1983, p. 468.

J. R. Partington, *Textbook of Inorganic Chemistry*, Macmillan, London, 1933.

J. Randles and P. Hough, *Spontaneous Human Combustion*, Robert Hale, London, 1992.

H. E. Roscoe and C. Schorlemmer, *Treatise on Chemistry*, vol. I: *The Non-Metallic Elements*, Macmillan, London, 1920.

The story of Holmes' Marine Life Protection Association Ltd is recounted in R. E. Threlfall, *100 Years of Phosphorus Making 1851–1951*, Albright & Wilson, Oldbury, 1951, p. 310.

There are numerous internet sites offering opinions and citing possible cases of spontaneous human combustion.

Appendix: The right chemistry

J. Emsley, *The Elements*, 3rd edn, Oxford University Press, Oxford, 1998.

N. N. Greenwood and A. Earnshaw, *Chemistry of the Elements*, 2nd edn, Butterworth Heinemann, Oxford, 1997.

Index